The Complete Pathway
to Ascension

*Thank you for supporting independent publishing
and the Mardukite Systemology Society*

MARDUKITE ACADEMY WORKBOOK EDITION

The Complete Pathway To Ascension

by Joshua Free

New Standard Systemology Professional Course

THE JOSHUA FREE IMPRINT
JFI PUBLICATIONS

© 2025, JOSHUA FREE

ISBN : 978-1-961509-61-0

Mardukite Research Library Catalogue No. "Liber-5+6" (5A+5B)
The Complete New Standard Systemology Professional Course

A MARDUKITE ACADEMY PUBLICATION
Mardukite Esoteric Library Grade-VI Reference

cum superiorum privilegio veniaque

WORKBOOK EDITION
First Printing—February 2025

Published from
The Joshua Free Imprint – JFI Publications
Mardukite Borsippa HQ, San Luis Valley, Colorado
Representing Mardukite Truth Seeker Press
Mardukite Academy of Systemology
and Founding Church of Mardukite Zuism

mardukite.com

Chart Your Flight For Ascension...
Then Let Your Spirit Fly!
"Zero" to "Infinity" at Lightning-Fast Speed!

Unlock your ultimate spiritual potential by removing the barriers to your true native state. Learn how to easily attain *Self-Actualization* and even help actualize others along the way. A greater appreciation and understanding of *Spiritual Life* and *Existence* awaits!

Each progressive lesson provides simple exercises and techniques that directly apply metahuman philosophy to increase your true Knowingness, improve your abilities in this life, and gain the power to decide and shape the destiny you'll experience in your next one!

This paperback workbook includes *all sixteen* of the original lesson-booklets (also available individually) for the complete *Pathway to Ascension* professional course.

Now in a single Academy Edition for the first time. Increasing Awareness; Escaping Spirit-Traps; Conquest of Illusion; Spiritual Implants; Games and Universes; Lifting the Veils; Spiritual Machinery; and many more (including *Systemology Processing Levels 0 to 6*).

It's time to discover who you really are... because you were never "Human."

This special premiere workbook edition is expanded to include: "Fundamentals of Systemology Basic Lesson-6" (as an introduction) and "The Complete Systemology Technical Dictionary (Version 5.0)"

∞

EDITOR'S NOTE

"The Self does not actualize Awareness
past a point not understood."
—*Tablets of Destiny*

This book contains a collection of materials from all
sixteen original lesson-booklets developed by Joshua Free
for the "Pathway to Ascension" Professional Course.
It is designed for Seekers that have already completed the
"Fundamentals of Systemology" Basic Course.

If you read an unfamiliar term not defined in the text,
refer to the "Systemology Dictionary" in the appendix.
It is also helpful to keep a quality dictionary nearby.

A clear understanding of this material is critical for
achieving actual realizations and personal benefit from applying
our philosophy as spiritual technology.

The *Seeker* should be especially certain not to simply "read
through" this book without attaining proper comprehension as
"knowledge." Even when the information continues to be
"interesting"—if at any point you find yourself feeling lost or
confused while reading, trace your steps back. Return to the
point of misunderstanding and go through it again.

Take nothing within this book on faith.
Apply the information directly to your life.

Decide for yourself.

∞

MARDUKITE ACADEMY OF SYSTEMOLOGY "THE PATHWAY TO ASCENSION" NEW STANDARD PROFESSIONAL COURSE

"Years ago, we realized that '*The Way Out*' would systematically resemble the routes by which we descended. We understood that the '*Gates*' reflected in our most archaic esoteric lore were pointing toward a *realization* that had been lost in translation along the way—and that our only hope of finding a *Map* to this *Pathway* was in recovering that lost understanding. I believe that our Systemology is successfully delivering a communication that is unparalleled in today's society—and most of you here can attest that: above and beyond the former gradients of knowledge available to us in our world, this work we are doing now is our best chance at 'making the grade' to reach our *Metahuman Ascension* in *this* lifetime; and for the first time in a very long time, reclaim the true power of the *Alpha-Spirit* and the freedom to experience an existence of our own true *Self-determined* creation. "

—*Joshua Free, <u>Backtrack Lectures</u>*

NEW STANDARD SYSTEMOLOGY PATHWAY TO ASCENSION PROFESSIONAL COURSE

The complete professional course for
The Pathway to Ascension,
collecting all sixteen lesson booklets
in one paperback workbook edition.

WELCOME, SEEKER!
LET'S CHART YOUR JOURNEY ON THE PATHWAY

Systemology is a "holistic" approach to understanding the human experience. It is not actually a singular "subject" in itself, but rather, a new way in which to view the many subjects of *Life* and all *Existence*.

This is a professional course in *Systemology*—specifically, how to *apply* the spiritual philosophy of *Mardukite Systemology* as a personal *"Pathway" to Ascension*. Our *Systemology* is a new approach to *"Self-Actualization."* It is completely relevant for the modern age and the future; and quite different from any previous similar attempts, or other traditions, you might find. What's more: it is applicable to anyone with any background.

This *"Professional Course"* series of lessons immediately follows the material given in the *"Basic Course"* series—available as six separate booklets, or in a single volume: *"Fundamentals of Systemology."*

This is a *new* presentation of *Systemology*, emphasizing the application of our philosophy for those *Seekers* that are *"Flying-Solo"*—or else working through their studies and exercises as solitary practitioners. This is a new innovation for *Systemology*.

To receive the greatest benefit from this study: it is expected that a *Seeker* will already be familiar with the fundamental concepts and terminology (previously relayed in the *Basic Course*) before using lessons from the *Professional Course*. This will allow us to cover the extensive territory of the *Pathway* much more quickly. [A *"dictionary"* of vocabulary is also provided in the *"appendix."*]

A NEW VIEW OF THE HUMAN SPIRIT

Systemology is not a religion and does not require any type of *faith*. It is, however, built upon a "spiritual" premise—and as such is an "applied spiritual philosophy." It is based on ancient teachings that we are *Spiritual Beings* essentially "wearing" bodies like clothes—or using them as "vehicles." Yet our true native nature is not *physical*, but beyond this existence; and we can certainly operate a "body" while maintaining an *Awareness* that is *outside* of it.

We are **all** *Spiritual Beings*—each of us a *unit* of *Spiritual Awareness*—that have experienced a very long *Spiritual Timeline* of existence. Although we might be particularly attached to the familiar "physical shells" associated with *this* lifetime, our true "*Spiritual Lifetime*" is seemingly *eternal*. We have been many things before *Human*, and we go onward as a *Spiritual Being* after our "*genetic vehicle*" of *this* incarnation perishes.

While a "spiritual" view of the *Human Condition* may not seem unique to our philosophy, just how often is the concept treated *systematically*? For that matter: just how many people, supposedly raised to this or that religion, or professing to believe one thing or another, actually live their lives as though they are *Spirits*?

As *Spiritual Beings* of immortal existence and infinite potential, we are not simply the "*creations*" of an even greater *Beingness*; we are, in fact, an integral part of that "*creative force*" which permeates all existence.

Our basic nature is to be a "*creative being*"—our highest goals are "*to create.*" And as such a being—which we refer to as an *Alpha-Spirit* in *Systemology*—we have run into some difficulties along the course of our *Spiritual Timeline* and found ourselves trapped within material *Universes* of our own collaborative *creation.*

Since we did not start out our existence in a trapped condition, it is correct to say that we have "*fallen*" from our native "*godlike*" states. It did not happen all at one, but progressively and systematically. We know our "troubles" have resulted from accumulated "barriers" and "blockages"—or *fragmentation*—during our vast experiences as *Spiritual Beings*. They are not because we lack something; but because of what's been added.

In *Systemology*, we systematically examine those routes by which we must have descended to reach our present condition, then reverse the direction of travel and chart a personal "*Pathway to Ascension.*" Of course, the exact "details" of the *Spiritual Timeline* will be different for each individual *Seeker*. However, we have been able to systematically chart our *Pathway* based on common patterns of *Human fragmentation*.

In the most basic terms: the *fragmentation* that defines our "downward spiral" consists of decisions or considerations where we deny our true nature. This includes those decisions to "*withdraw*" rather than "*reach*"; where we choose to *not-know* rather than *know*; to *not-communicate* rather than *communicate*; and ultimately, to take *no-responsibility* for being a *creative-cause*, and therefore succumb to being an *effect.*

But there is *hope!* And much more importantly: there is an effectively workable *way out* of the mazes and traps of our existence. If you are reading this now, you have already begun to gather your tools and build up the "*horsepower*" necessary to break the gravity holding your *Spiritual Beingness* to the *Human Condition.*

STUDYING THE PROFESSIONAL COURSE

Most *Seekers* study and practice *Systemology* at-a-distance and independent of the "Mardukite Academy." This means that the *books* (and to a lesser degree, the *internet*) are the only means of direct contact a *Seeker* maintains with the "Systemology Society" during their studies. A continuing *Seeker* from the "*Basic Course*" will be familiar with the style of study found in *this* course.

Misunderstood words are the most common reason an individual abandons studying a subject. When a misunderstanding occurs, *Awareness* declines. These misunderstandings start to "stack up" after the first occurrence, and as a result, the level of interest and attention will also decline. This is how a "confusion" develops; and the individual will get "bored" with the subject, feel tired, and unable to concentrate.

One solution is to return to the part of the material that was still interesting and enjoyable to read. When scanning around that area of text, there is likely to be a new word (or new specific use of a familiar word) that is unclear, but was passed by unnoticed. All *Systemology* books include their own *glossary*. Using this *glossary* and a high-quality dictionary will help resolve this misunderstanding once it is located.

An effective education of any subject is taught on a *gradient*. This is what is intended by presenting the study of something as "*grades.*" Rather than treating a subject as one total mass, true learning is achieved by increasing one's understanding with a *gradual* increase upward. The *ascent* to a mountaintop is not successfully achieved in one leap, but by targeting and reaching specific checkpoints along the way.

This *Professional Course* consists of a series of lessons that gradually increase a *Seeker's* ability to understand and apply the practices and techniques of *Systemology* as a complete "*Pathway to Ascension.*"

Each lesson of the *Professional Course* applies *Systemology* to a particular subject (or focus). It is best if the entire course can be studied and applied in sequential order. These lessons also employ a style of practice or technique called "*Systematic Processing.*" [An introduction to this methodology is provided later in this workbook as excerpted from the *Basic Course.*]

To study the *Professional Course* just like a student at the Academy: a *Seeker* reads through all instructional material and applies each exercise (or "*process*") presented in the text to the extent they comfortably can, before continuing on to the next lesson.

When first starting on the *Pathway* as a *Solo* practitioner, without the aid of an experienced *Pilot*, a *Seeker* shouldn't "push too hard" or allow themselves to get too "stuck" on any one area (lesson) or *process*. It is not expected that any one area will be completely handled when first introduced. For optimum results, it is expected that a serious *Seeker* will make more than one "pass" through the entire *Professional Course.*

The *Professional Course* is not altogether different from other forms of practical or technical education: where the instruction and exercises are delivered to a completion, and then a student further increases their abilities, strength and skill-level by applying additional practice throughout their life. Therefore, a student should not concern themselves with perfectly mastering each step (or lesson) before progressing forward.

Additional passes through the material are likely to result in different "*realizations*" (an increased *level of understanding*) than a previous time. New "layers" of *Knowingness* may now be accessible during a *process* that may not have been before. It is important to avoid invalidating the progress you've made just because one area is not completely handled right away, or if a certain *process* seems too difficult on the first pass.

CHARTING A COURSE ON THE PATHWAY

Although we can communicate a systematic structure to *fragment-ation,* the personal journey experienced along the *Pathway* will be different for each *Seeker.* For example, certain areas will seem more *"turbulent"* or difficult for one *Seeker* than another. We tend to say that these areas have more *"charge"* on them—or that they are more *"heavily charged."* It is best to handle such areas when you are already feeling "good" and not in a situation (or condition) where that specific area is consistently being *"triggered"* or *"restimulated."*

As an applied philosophy, *Systemology* "theory" can be easily utilized in the "laboratory" of the "world-at-large" in everyday life. This is implied within the basic instruction of each lesson. Unlike other "sciences" that conduct experiments by making a change to some "objective variable" *out there* and waiting to see an effect, our focus is the individual (or *Observer*) themselves, and how *they* affect the *"Reality"* perceived.

In addition to applying *Systemology* "New Thought" to everyday life, our philosophy is applied by using specific exercises and systematic techniques. These *"processes"* provide the most stable personal gain (and *realizations*) for each area; but only when actually applied with a *Seeker's* full *"presence"* and *Awareness.*

This *Professional Course* is designed so that it may be easily read and studied with little concern for what "dangers" these teachings —or *processing*—might unleash. However, there are still some guidelines that pertain to the "best-uses" of these course lessons, particularly if a *Seeker* intends for stable development.

Skipping over too much material/*processing* in early lessons may make attempts to understand (or apply) later lessons more difficult. However, once the complete *Professional Course* is worked through at least once in its entirety, specific areas can then be later returned to and treated with a greater sense of *Awareness* and *"presence"* than before. Of course, in *"Traditional Piloting,"* the rate of processing is monitored by an experienced practitioner; but in "*Solo-Processing*," a *Seeker* must regulate their own progress on the *Pathway*.

Applying a systematic technique is called "*running a process.*" The *processes* are designed with very simple instructions or "*command-lines.*" To *run* a *processing command-line*, a *Seeker* may be assisted by the communication of that *line* from a "*Co-Pilot*" (as in "*Traditional Piloting*"). But even then, a *Seeker* must still personally "input" the *command* as *Self.* For this reason—and quite thankfully —*Solo-Processing* is possible.

TAKING FLIGHT ON THE PATHWAY

Processing Techniques are intended to treat the *Spiritual Being* or *Alpha-Spirit*; the individual themselves. It is applied by the *Alpha-Spirit*—then *Self-directed* to the "Mind-System" or even a "body" (*genetic-vehicle*), both of which are "constructs" that the *Alpha-Spirit* (*Self,* or the "I-AM" *Awareness unit*) operates, but neither of which is actually *Self. Fragmentation* causes *Humans* to falsely identify *Self* as the "*Mind*" or even a "*Body.*"

The *Professional Course* lessons are designed for the *Beginning Seeker* in mind—one that may have an understanding of theory, but with little experience in practice. That being said: each of these lessons may be used toward total *Beta-Defragmentation* within a specific area. There are also more *processes* given for each subject than may be necessary to achieve an *ultimate end-point realization* on that entire area.

Some *processes* can be treated quite lightly at first; others may require a bit of working at in order to get "*running*" well. It is important to set aside a period of time when you can be dedicated to your studies and *processing.* This period of time is referred to as a "*processing session.*" The reason for this, is that when a *process* does start *running* well, it is important to be able to complete it to a satisfactory "*end-point.*"

The purpose of *systematic processing* is to be able to really "look" at things and even determine the *considerations* we have made—or attitudes we have decided—about *Reality* as a result of those experiences. It doesn't do us much good to simply "glance"—or to *restimulate* something uncomfortable and then quickly *withdraw* from it once again, leaving more of our *attention* yet again behind and held fixedly on it.

Generally speaking, a *Seeker* continues to *run* a *process* so long as something is "happening"—which is to say, the *process* is still producing a change. Usually this is evident by the type of "answers" that a *command-line* helps a *Seeker* originate from the database of their own *Mind-System*. The *command-lines* do not "do" anything on their own. They assist a *Seeker* to direct their own attention toward increasing *Awareness*.

Of course, a *Seeker* may also cease to generate new "data" from a *process* without reaching an *"ultimate" realization* as an *"end-point."* It is possible that additional "layers" (or even other "areas") require handling before anything "deeper" is accessible. If this is the case, end the *process*. But, if a *Seeker* is *withdrawing* from something uncomfortable that was incited or stirred up, then a *process* is *run* until they feel "good" about it.

In case the thought of encountering *"turbulence"* is a concern: the techniques given as *"Opening Procedures"* of a *Formal Session*, and those found in the earliest lessons of the *Professional Course*, are quite useful when applied as "safety nets" for maintaining *Awareness* and *presence*, even when *Flying-Solo*.

One of the benefits to *Flying-Solo* is that *processing* is entirely *Self-determined*. This already provides a certain built-in "safety" for a practitioner. Anything you *restimulate* by *Self-determinism* is *your thing*. It is not incited by external *other-determined* influences (or other "source-points" in existence) that make you an *effect*. It can be more easily handled in *processing*—or you can simply let things "cool down" and come back to it again.

While it may seem "mysterious" to beginners, a *Seeker* gets a sense for knowing how long to *run* a *process* only with practice. Once you have spent some time actually applying the *Professional Course*, there are many aspects that become "second nature" because they are, in fact, a part of our true original nature. All we have done is *"reverse engineer"* the routes of *creation* and *consideration* that are already *our own*.

AN INTRODUCTION TO PROCESSING (BASIC COURSE LESSON SIX)

PRACTICES OF SPIRITUAL AWAKENING

The philosophy of *Systemology* may be applied to many fields and areas of everyday life. When *applied* within our tradition, practices and exercises toward "spiritual awakening"—or the "*Pathway to Ascension*"—are referred to as "*systematic processing.*" Some of the inspiration behind this is derived from sources highlighted in the *Fundamentals of Systemology Basic Course* "*Lesson 5.*" But really this is the product of many research years and innumerable sources.

"*Systematic Processing*" is only *one* part of our philosophy. It is itself an entire "methodology" for applying *Systemology* as a personal practice of techniques and exercises. We call it "*systematic processing*" because it is a precise practice (or "ritual") that *knowingly* mirrors or duplicates the "systematic processes" of the *Mind*. It is too broad a topic to cover fully in this one lesson, but we can introduce its practice.

One of the basic goals of "*processing*" is to treat *knowingly* (at an analytical level of *awareness*) what is happening "automatically" or "compulsively"—or otherwise *unknowingly*. However, *processing* may be better understood as techniques aimed toward "spiritual development" and "ability enhancement." Of course, these *processes* or techniques are *applications* based on our *philosophy*. Hence: "applied philosophy."

Effectiveness of *systematic processing* is entirely based on our principle axiom that: The "I" or *Self* is an *Alpha-Spirit* operating from a *Spiritual ("Alpha") Existence* and employing a *Mind-System* to perceive the sensory experience of a *genetic-vehicle* (or 'body') that interacts in a *Physical ("Beta") Existence*. This axiom *is* "The" *Fundamental of Systemology*—as already explored throughout the previous lessons.

Systematic Processing does resemble some practices from our predecessors in history; perhaps because it reaches for some of the same goals and ideals. It is, however, *not* synonymous with (or equivalent to) these other methods—meditation, prayer, mental healing, therapy, psychoanalysis, *&tc.*—at least not as they are commonly understood by the general (*exoteric*) public. Therefore, we tend to avoid using such terminology.

The most common application of *Systemology* is "*defragmentation*"—but this is only *one* use of "*processing.*" *Defragmentation* is best understood with the model of "*knowns* versus *not-knowns*" given in the introduction. While many assume this means only a *knowledge* of "facts," the same example could illustrate other uses of *processing*, such as a gradual increase of "dormant" (or forgotten) "*spiritual ability*" into "*actual ability*" *&tc.*

BASIC METHODS OF PROCESSING

An "applied philosophy" is really only as effective as an individual *understands* that philosophy and is able to *apply* it in practice. If a *Seeker* is unable to *understand* the philosophy enough to *apply* it, the logical solution is to find an individual that is professionally trained to *understand* it, until the *Seeker* has a *reality* on it themselves. For this, we developed the idea of a "*Pilot*" that could assist in guiding a *Seeker* on the *Pathway*.

This "*Pilot*" concept led to three basic "methods" of *systematic processing*, originally distinguished as *Piloted, Co-Piloted* and *Flying-Solo.* Over the course of many years of additional work, the meaning implied by these terms has evolved slightly during development. But, to briefly describe, they are:

Piloted Processing—an untrained *Seeker* is processed by a professionally trained "*Pilot.*"

Co-Piloted Processing—two *Seekers* in training take turns processing one another; or a *Seeker* is processed by a trusted friend, reading from a book.

Flying-Solo—a *Seeker* in training processes themselves.

When first established in the 2010's, all *systematic processing* was intended to be "*Piloted*"—administered by a book-trained or Academy-trained "*Pilot*"—or "*Co-Piloted*" with a friend, or fellow *Seeker*, as "*Pilots-in-Training.*" And this is exactly how "*Route-1*" (the first experimental method of *processing*) is presented in *"Tablets of Destiny Revelation"* (2019). More recently, *"Co-Piloted"* seems to apply to anything not done *"Solo."*

Soon after our debut publication, we realized the obvious limitations of a *Piloted* approach. We decided that when developing additional routes, they would need to apply to both traditional *"Piloting"* and solitary practitioners (who were *"Flying Solo"*). That being said, there *are* some *processes* (such as *"Route-1"*) that greatly benefit from having another individual present; yet others are just as productive when "run" *Solo.*

A *Seeker* working alone has only the option of being their own *Solo Pilot.* As such, they are responsible for all the "training" as a *Pilot,* and also managing the *"processing session"* for themselves as a *Seeker.* A certain level of *Self-Determinism* and *Actualized Awareness* must already be in place in order to successfully *"Fly Solo,"* since no one else is *present* to help maintain a *Seeker's* "presence" (*attention* and *Awareness*) *in session.*

The phrase *"Self-Processing"* is sometimes used in place of *"Solo"*—but this is not necessarily the most accurate term. Whether a *Seeker* is practicing alone or not, all *systematic processing* techniques in *Systemology* are *"Self-processed"*—which is to say, they must be *processed* by *Self.* And if it seems to you that *Systemology is* sometimes talking about software programming, or operating a computer, you're not alone; but it works.

A *Pilot* may assist in directing attention—or "redirecting" attention, if it strays—but the "command line" (verbal instruction) of a *process* is not a "magic spell" that "does something" by itself. To be effective, the *Seeker* must actually *apply* the "command line" of the technique as *Self* to *Self.* This is what we mean by "running a process"—because essentially, a *Seeker* is *processing* the "command" and resulting *data* as *Self.*

SYSTEMATIC PROCESSING SESSIONS

New *"realizations"* and increased *Actualized Awareness* improve a *Seeker's* handling of everyday life. This may be enhanced through educational training, but a more effective means of reaching these higher ideal states is by combining book-learning with actual practice of the *Systemology* techniques and experiencing the exercises we refer to as "*systematic processing.*" A few of these have already appeared in the *Basic Course.*

To be strictly technical, the practice of *systematic processing* is conducted in a formal "*session.*" We prefer the term "session" because of the common mystical and/or religious connotations associated with the word "*ritual*"—or even "*meditation.*" Our use of the phrase "*processing session*" is most accurate since it implies a specific duration of uninterrupted time set aside to focus on a particular procedure—or *process.*

Rather than conceiving the idea of a "session" as being similar to some "hourly counseling" in some other tradition, a "session" is a period of time for "running a process" toward an intended result or "end-point." A single "session" could last twenty minutes or two hours or more (taking breaks as needed, of course).

Proper procedures for conducting a "session" are followed whether a *Seeker* is *Solo* or *Co-Piloted.* This allows a *Seeker* to treat *systematic processing* as its own unique activity. For example: "*bathing*" is its own unique activity; but it also carries with it a whole regimen of "steps" that are followed in sequence—and which eventually become "*routine*" to the *process.* A *Seeker* practices to achieve that same level of familiarity.

That being said: even the fundamental practice of "starting a session," providing "*presence*" (*attention* and *Awareness*) to be "in session," then formally "ending the session," is a *systematic procedure* or *routine* in itself. To ensure effective processing, we add a step to getting "*in session.*" A *Seeker* "scans through daily life" for upsets or problems that are "holding" parts of their *attention* or *Awareness*, even if only on the periphery.

When *things* are not handled analytically—or "above the surface"—they remain in suspension. A *thing* does not *disappear* by our withdrawing from it; it waits around, albeit out of view, to be seen "*As-It-Is.*" Many aspects of living out the *Human Condition* have a tendency to keep our *attention* suspended. Even when we go about performing other actions, the totality of our available *Awareness* is not generally present.

For example: we might be preparing a meal for our family, but at the same time we are thinking about problems at work, difficulties in managing the bills, and the driver that almost collided with us on the way home. It is also at these moments, when not living deliberately, that accidents can occur—because, the individual "isn't there" or their "mind isn't on it." Either way, they have withdrawn part of their "*presence.*"

Applying "*presence*" to a session is one of the most critical fundamentals of *processing*. Without it, a *Seeker* is not actually "*present*" to be *processed*. We aren't interested in *processing* a "body" or "computer-mind"; we are communicating directly with *Self*, the *Alpha-Spirit* that *utilizes* a "*Mind*" and "*Body*"—and *Self* maintains a "spiritual identity" that is independent of a "*Mind*" and "*Body.*" It is *Self* that we want *present* in session.

Systematic Processing operates by taking *knowing control* of an individual's available *Actualized Awareness* and then increasing it—both the level of *control* and the available *Awareness*. This is similar to taking a small amount of *certainty* or *ability* and "building up" from there. For a *processing session* to be effective, a *Seeker* cannot be withdrawing their "*presence*"—distracted by the upsets of daily life. This must be handled first.

It is no great mystery that having part of our *attention* on all of the aspects of living (that we hold at length on the peripheral boundary of our conscious *Awareness*) might inhibit our focused concentration or the achievement (or experience) of other "states." But, we can't simply ignore this fact, telling our *Seekers* and readers that we hope they feel better one day and come back to us when they do. We handle it *in-session*.

HANDLING IN-SESSION PRESENCE

Establishing "*in-session presence*" is not simply a preliminary step; it is a *systematic procedure* in itself. In fact, it is actually the *single* element underlying all other practices—of meditation, prayer, ritualism, &tc.—that produces any real *effects*. Of course, in other practices, these *effects* are misappropriated to some other *cause*.

The "location"—the environment or "setting"—is one of the first things to consider for establishing a session. This may be in an un-interrupted space outdoors, or in a quiet room. On occasion, the instruction or "command line" of a *process* may be directed toward a particular environmental focus—such as an "item" *in the room*, or a "person" engaged with others in a *public place.* These are indicat-ors of the intended *setting.*

With the *setting* identified, the next step is making certain a *Seeker* is comfortable and relaxed within that environment. It is unpro-ductive to attempt *processing* in an environment that is actively a source of "turbulence"—or worse, the very type of turbulence that is going to be *processed.* There is no reason for some cult-like dis-connection; but a *Seeker* should have a "*retreat*" available to them while regaining balance.

Once a "safe location" is available, our next concern is the *Seeker's* comfort and familiarity with the material and intended practice. In traditional *Piloting,* this would include not only a confidence in the philosophy, but also the individual ("*Pilot*") administering it. When applied to a *Seeker* that is not also in-training, the total con-trol of *processing* must be handled by a *Pilot* until the *Seeker* as-sumes responsibility for that control.

All of these factors boil down to what we consider the first ele-ments of "*in-session presence*" that a *Seeker* provides to their exer-cises (*processing*)—the *attention* that they have available and are willing to provide to the session *in* present-time. And when we say "*presence,*" we mean, quite literally, the *Awareness* of actually "*be-ing present*" in the session. This "*presence*" *is* the unifying factor of all effective spiritual practices.

Other factors being considered, the actual "orientation" of a *Seeker's presence*—present-time *Awareness*—*with* the present space-time *setting* or environment, is the primary "opening procedure" for a *processing session.* An integral part of this procedure is de-termining if there are any aspects from daily life that are *inhibit-ing presence.* Sometimes the handling of these aspects alone is enough to qualify a complete session.

A *Seeker* handles the upsets or energetic-turbulence that is already present—or in stimulation—on the *ZU-Line* before attempting to *resurface* or reveal additional layers of *fragmentation.* Energy is always relayed as a "communication" (even internally for our "*per-ceptions*"); therefore, most upsets can be categorized as a type of "communication-breakdown" or "break in reality"—an unexpected interruption in the energy *flow*.

Different *processing* techniques target specific aspects of life. But, as a general rule—and in regards to *presence*—the basic idea is to *confront* ("come around to face") or *reach* rather than *avoid* or *withdraw.* Those *things* that we have "blanked out"—or choose not to look at—are still "energetic masses" surrounding our "field" of *Awareness.* All a *Seeker* really needs to do is "*acknowledge*" they exist so *attention* can shift off them.

To approach this *systematically*, a *Seeker* "spots" whatever incident or aspect is bothering them, then *looks* it over carefully and *analytically*. This may be done as thoroughly as is needed—noticing various things about it, but essentially *acknowledging* or *confronting* a thing "*As-It-Is.*" This allows a *Seeker* to put some distance between themselves and the energetic-mass ("*problem*") rather than treating it as though it were "present."

We return to our previous example of worrying about things while preparing a meal, even if *unknowingly*, which leads to an accident. What is taking place is that the individual is treating those other things as though they are *present* in the environment—giving them "attention-units" as though they are an imminent threat. Of course, by agreement, they are *real things, real incidents*, but *are* they actually "*present*" *in session*?

Once a troublesome aspect or incident is analyzed (as above), there should be some feeling of relief and an ability to refocus *attention* and *presence in-session.* A "problem" should *seem* "further" away rather than overwhelmingly close. If this is not the case, there may be another upset or distraction. If the troublesome aspect *seems* closer, then alternate *looking* at something in the incident and something in your environment.

This brings us to the other part of establishing *presence in-session*, which is "orientation with the present space-time" (or environment). Essentially, this is a part of what a *Seeker* is employing when *attention* is alternated by *looking* at something in the incident or situation and then something in the physical surroundings. We employ a similar technique as an *"opening procedure"* for all formal sessions as well.

In the case of handling "problems" or upsets, the alternation allows a *Seeker* to "unfix" *unknowing attention* on something that is being "fixedly" treated as "present" (*in-the-now*) when it is really *somewhere* else or a *past* situation no longer happening presently. These aspects that *attention* is *unknowingly* "fixed" to are what reduces available *Actualized Awareness* that is applied to any "present-time" activity—including *processing.*

Once *Awareness* is able to be focused or concentrated, this additional step—"orientation with present space-time" is essentially exactly what it sounds like: "orienting" *Self* in the present time and space of the session. This means bringing total available *Awareness* to the present-time location (space) of the session and the control of the "body" *in-session.* With regular practice these steps take less time further along the *Pathway.*

The experience of *"Presence In-Session"* will be familiar to continuing students and readers. Versions of sample techniques used as "opening procedures" of a *"Formal Session"* are given in earlier *Basic Course* lessons as follows:

Lesson 1, Exercise 1 and 2;

Lesson 2, Exercise 1 and 7;

Lesson 3, Exercise 1, 2 and 7.

THE FORMAL SESSION

All *systematic processing* is practiced as part of a *"formal"* systematic *processing session.* Of course, there are some "exercises"—such as those included at the end of each lesson—that are effective even when practiced on their own. However, whether *Solo* or *Co-Piloted,* a *Seeker* applying our methods as a total *Pathway* toward *Spiritual Awakening* (or *"Ascension"*), benefits most by practicing techniques *systematically.*

A session is considered *"formal"* because it follows a specific pattern or ceremonial formula of *formal* action and communication. There is a *formal* "beginning" and "ending" of a *systematic session.*

Several *processes* may occur within a single session; each one "started" and "stopped" (in turn) as a *formal* act.

Traditional *Piloted Processing* differs from *Flying-Solo*, but the basic formula for a "*Formal Session*" is the same. In *Piloted* (or *Co-Piloted*) *processing*, the factor of *communication* between a *Pilot* and *Seeker* must be handled in addition to the *session* and *processes*. In *Solo-Processing*, the *communication* of a *process in-session* is all handled "internally"—between *Self* and the "*Mind*" (and "*Body*") without being directed by a "*Co-Pilot*."

A precise instruction for a *process* is referred to as a "*processing command line*" (or "PCL"). This is named for the act of "inputting" a "command line" into a "computing device." In traditional *Piloted* sessions, a *Seeker* receives a communicated "command line" from the *Pilot*, then communicates the "command" to the *Mind* as *Self*. *Solo* or not, a *Seeker* directs their own *Mind* to *process* the "command"; another *Pilot* only assists this.

A "*processing command line*" (PCL), by itself, is *not* a "magical incantation" that spontaneously produces *actualization*. A misunderstood one, however, *does* have the "power" to slow or stop forward progress. A *Seeker* that is studying/training on their own usually does not have an issue *in-session*, because they can get a clear understanding beforehand. A *Pilot* is responsible for making sure of this for a *Seeker* not in-training.

For example: if you take someone randomly off the street and ask, "*would it be okay if we start this session?*" Well, there is obviously little context there. Even an individual that is interested in getting *processed* might not understand the use of the word "session" at first unless some of the basic *Systemology* philosophy is explained. A *Solo-Pilot* simply accepts total responsibility for attaining this understanding for themselves.

Solo-Sessions may be run "*silently*" as direct "mental commands"—but this is not an absolute rule. In either case, a *Seeker* should be focused on handling the commands directly and not *imagine* also *being* "another person" that is giving themselves the commands. Such "add-ons" are unnecessary and counter-productive. However, the same formal session "script" that a *Co-Pilot* communicates, a *Solo-Pilot* reads and *Self-Directs*.

The whole purpose of a *processing session* surrounds the idea of an individual focusing and increasing their *Self-directed* control—or *Self-determinism*. In order to retain this focus, every *one* PCL of a formal session or *process* must have a *Seeker's* total *presence*. This level of focused direction is one benefit of traditional *Piloting*—but in *Solo*, a *Seeker* is instructed to keep a piece of paper over portions of a script/process not yet used.

What follows is a basic script from a *Formal Session* used for training purposes at the Academy. It is a guideline only—based on a transcript of a traditional *Piloted Processing Session*. It is, however, easily adapted to use for *Flying-Solo* once a *Seeker* has practiced the exercises used for achieving *presence in-session*.

1. <u>BEGINNING THE SESSION</u>

"Would it be okay with you if we begin this session now?"

"Okay."

"Start of session."

2. <u>OPENING PROCEDURES</u>

 A. Presence In-Session

"Is there anything going on that might keep your attention from being present in-session?"

 (if *"no,"* acknowledge and go to B.; if *"yes,"* continue below)

"Okay. Tell me about it."

"Alright. How does that problem seem to you now?"

 (if *"further away"* or handled, acknowledge and go to B.; if *"closer"* or more turbulent, continue below)

"Spot something in the incident; Spot something in the room."

 (this alternating command line is repeated as needed)

 B. Orientation in Present Space-Time

"Get the sense of you making that body sit in that chair."

"Okay. Get a sense of the floor beneath your feet."

"Do you have that real good?"

(if *"no,"* acknowledge and repeat A.; if *"yes,"* continue below)

"Recall a time something seemed real to you."

"Tell me something you notice about it."

"Look around and spot something in the room."

"What do you notice about that?"

(these last four command lines are repeated in series as needed; acknowledge and continue below)

C. Control of Body and Mind In-Session

(two dissimilar objects—here given as *"Item-1"* and *"Item-2"*—are presented and placed within reach; or alternatively, at two distant points in the room, in which a command line for "walking" between them would be inserted)

"Pick up Item-1."

"Tell me about its weight."

"Tell me about its color."

"Tell me about its texture."

"Put it down."

"Pick up Item-2."

"Tell me about its weight."

"Tell me about its color."

"Tell me about its texture."

"Put it down."

(this series of command-lines may be repeated several times; when there is no communication-lag for several full series, and duplicate answers are reoccurring, acknowledge and continue below)

"Choose an object. Decide when you are going to reach for it. Then make that body pick it up."

"Now decide when you are going to put it down. Then make that body put it back where it was."

(repeat as needed; when there is no communication-lag for a full series of command lines, acknowledge and continue below)

"Close your eyes. Put all of your attention on the upper two back corners of the room and just get real interested in them for a while."

(if there are no visible signs of "strain" after two minutes, acknowledge and continue below)

D. Establishing the Session

"Do you have any goals for this session, or anything in particular you want to address?"

(acknowledge, then start a process)

3. <u>STARTING A PROCESS</u>

"I would like to start a process; would that be okay?"

"Alright. The command lines are ---. Does this make sense?"

(if "*no*," clear up any misunderstood words; if "*yes*," start the process)

4. <u>CHANGING A PROCESS</u>

(only the wording in a command line may be changed to make it more workable for a *Seeker*; to change processes altogether, the present process must reach an end-point)

Example: a Seeker expresses inability to "imagine" or visualize imagery.

"Okay. Well, just 'get a sense' of..." or *"Just 'get the idea' of..."*

Example: a Seeker expresses discomfort (or withdrawal from) recalling a particular incident.

"That's fine. What part of that incident 'could' you confront?"

5. <u>STOPPING A PROCESS</u>

(when an end-point has been reached on a repetitive-style process)

"We'll just run this process a couple more times if that's okay with you?"

(general process is run two more times)

"Okay. Is there anything you would like to tell me before we end this process?"

(**or**, if an end-point "realization" is communicated from a process)

"Alright. Very good."

(the formal end of a particular process requires a command-line)

"End of process."

6. <u>ENDING THE SESSION</u>

(once a process, or series of processes, is completed)

"Is there anything you would like to tell me before we end this session?"

(if *"yes,"* acknowledge and handle it with communication before ending the session; if *"no,"* continue below)

"Would it be okay if we ended this session now?"

"Okay."

"End of session."

THE BETA-AWARENESS SCALE

    4.0 SELF-HONESTY (BETA)
    3.9 "Vibrant" ("Charismatic")
    3.8 "Enthusiastic" ("In Love")
    3.7 "Energetic"
    3.6 "Cheerful"

3.5 CONFIDENT ("Positive")
3.4 "Determined"
3.3 "Eager"
3.2 "Alert" ("Attentive")
3.1 "Strong Interest"
3.0 INTERESTED ("Content")
2.9 "Small Interest"
2.8 "Encouraged"
2.7 "Disinterest"
2.6 "Doubtful"
2.5 INDIFFERENT ("Tolerant")
2.4 "Bored"
2.3 "Dislike" ("Neglectful")
2.2 "Tired"
2.1 "Monotony"
2.0 INVALIDATING ("Pessimistic")
1.9 "Antagonism"
1.8 "Pain"
1.7 "Confrontational"
1.6 "Violent"
1.5 ANGER ("Negative")
1.4 "Hateful"
1.3 "Spiteful"
1.2 "Resentment"
1.1 "Anxiety"
1.0 FEAR ("Afraid")
0.9 "Terror"
0.8 "Numb"
0.7 "Evasive"
0.6 "Loss"
0.5 GRIEF ("Sadness")
0.4 "Depression"
0.3 "Victimization"
0.2 "Hopelessness"
0.1 "Apathy" ("Unconsciousness")
0.0 BETA CONTINUITY (Organic Death)

LESSON ONE
"INCREASING AWARENESS"

FIRST FLIGHTS ON THE PATHWAY

"Processes" are systematic techniques—or *actions*—that are repeated toward a specific *end-point* or result. In most cases, they consist of a repetitive instruction—or *"command-line"*—that is *run* over and over until something happens. This is what produces a *realization*, which may not happen the first few *runs*. The same idea applies to hammering a nail with multiple strikes rather than simply pushing hard against it.

The first lessons of the *Professional Course* are in *some* ways a review of many of the *techniques* introduced in the *Basic Course*. However, in this course, we will examine them much further as *processing* applications. In the *Basic Course*, just a few of the most critical *processes* are given as light *exercises* to supplement introductory lessons on the fundamental theory and philosophy of *Systemology*.

Much of the basic theory behind *"Systematic Processing"* may be found in *Lesson-6* of the *Basic Course*. In this *Professional Course*, we will be handling the *processes* directly as applications of our philosophy—and as a practical approach to unfolding the essential map of the *Pathway*, as researched by the *Systemology Society*.

SELF-DIRECTING ATTENTION

Consider for a moment that an *Alpha-Spirit* is able to *Self-direct* its nearly unlimited *Awareness*—and this has taken place across a long span of perceived existence (that we refer to as a *Spiritual*

Timeline). Although the potential *Awareness* seems without limit, the ability to *handle* it has actually deteriorated over time, and with it, the potential "considerations" an *Alpha-Spirit* maintains of its own *Beingness*.

Understand that the original potential is *not* truly lost to us. It has, however, become *fragmented*. And by this, we mean that an individual's *attention* gets drawn toward painful instances and dangerous circumstances—and other "human problems" that can "fix" our *attention*. This, in fact, lowers the total *Awareness* immediately available to us; it affects just how much of our true *Self-determinism* is "in play."

Systematic Processing, in general, is an exercise in "*selectively directing attention.*" This is one of the reasons it is so effective for increasing *Awareness*. The exercises given in the *Basic Course*—particularly those related to "*presence in-session*" or the *Opening Procedures* for a *Formal Session*—are the most fundamental *processes* by themselves, because they "orient" a *Seeker* in present time and space to make any other *processing* workable.

Our *processing* methods are effective when they can collect (or concentrate) a *Seeker's* available *Awareness* (or "actualized" *Awareness*) and then increase it. This is essentially the *opposite* of hypnotism. In *processing*, a *Seeker* "frees up" more of their available "*attention units*" by taking them off of whatever they have been fixed on *unknowingly* throughout one's existence. These "*units*" have gotten stuck on things along the way.

"*Fragmentation*" is an archaic systemological term we still use today because it implies a "dispersal" of energy—or quite literally the "fracture" of a wholeness or totality into many parts. It is meant that some *thing* is in the way of a "clear view" (or "clear communication").

In most cases, *fragmentation* concerns what we don't want to confront directly—so we kind of "shut down" on those areas and withdraw, but without actually taking all of our *attention units* off of it. We don't really want to deal with it, but we can't trust it not to "bite" us when we're not looking. This "area" sinks into the shadowy gray parts of our *Awareness* until it becomes completely handled *unknowingly* "on automatic."

We have also retained the word "*imprint.*" This is best understood in this wise: you may have noticed that you are likely to give something more *attention* when it is first encountered—and certainly, common language makes frequent use of the phrase "*first impression.*" It is at these instances that we essentially take the data we have received and duplicate it as our own *Reality*. And this is what we agree to as *being Reality*.

When an *Alpha-Spirit* stops "*looking*" and "*creating,*" and starts using *imprinting* as the basis of *Reality*, the total available *Awareness* declines. The individual is still carrying the same amount of *Spiritual Awareness Energy* (or *ZU*) as they always have—but these energy stores have become increasingly "solidified." The heavier or more solid these energy units are, the farther below the line of *Actualized Awareness* they sink.

For *processing* to be effective, we begin with those *techniques* that will bring together those "*attention units*" that *are* actually accessible to a *Seeker*. These are also useful as general methods for "*selectively directing attention*" on other tasks, or in times of mental or emotional strain. "Orientation in Present Space-Time" is also a critical part of the *Standard Opening Procedures* for a *Formal Session*.

The alternating "*command-lines*" in the sample script for a *Formal Session* (as given in the *Basic Course*), are:

"*Look around and spot something in the room.*"
"*What do you notice about that?*"

Of course, this is taken from a "*Traditional Piloting*" transcript, which involves two individuals—a *Pilot* and a *Seeker*—and is dependent on their relay of communication. Many variations of this are effective. An alternative *processing command-line* (or "PCL") that may be more applicable for *"Solo-Processing"* (rather than the communicative approach), is:

"Look around and notice things. Locate precise points on the object, moving quickly from one point to another."

In basic terms, this *knowingly* duplicates the original basic systematic process of *imprinting*. This is what we do when we encounter a new person or enter a new place—at least *before* we tend to leave our ability to *perceive* and *create* (or *duplicate*) on automatic. We take a permanent snapshot to base our total *Reality*, but its vibrancy often fades. This is why the vividness of *Life* and the *Universe* can seem so "dull" sometimes.

Now, it is important when doing this *process* that you are actually "spotting precise points" with your full *attention* and not just casually glancing all about. Remember that these *are* "systematic processes" in spite of how plain the language used actually is, or how simple or trivial the action required may seem.

When using this *process* during periods of emotional turbulence or mental strain, you may suddenly feel more alert or clearer in your perceptions. Even if already awake and alert, there should be a sense of improvement, or perhaps the room may seem to be a little "brighter" than before. In either case, you would acknowledge (even to yourself, if *Flying-Solo*) that the *process* has reached a satisfactory *"end-point."*

In addition to the improved "orientation in present space-time," this *process* also demonstrates the ability of the *Alpha-Spirit* to direct its *attention* and therefore control its experience of a mental state. Such is an example of a potential *"realization"* that might also spontaneously occur as a result of *running* this *process*.

ADVANCED APPLICATIONS

There are many applications for the previous *process* other than focusing *presence* for additional *processing*. In fact, early experiments with *"presence"* led to an entire route of advanced work (otherwise referred to as "A.T.") regarding *perception* of—and *operation* in—existence as an *Alpha-Spirit*, but independent of *any* body. Such is the *true* native state of *Self*.

This more advanced subject is presently brought up because of its ability to illustrate just how "not-trivial" the previous *process* really is. This became evident when *Seekers* assisting with research at the *Systemology Society* began to experiment with the previous *process*, but with their *eyes closed*.

Whether eyes are open or closed, this type of technique is called *"objective processing"* because it pertains directly to the "objective environment" or *Physical Universe* (*Beta-Existence*). This is quite different from a *"subjective process"* that calls for a *Seeker* to "remember" or "consider" something. There are other types, but the majority fall within either of these two categories.

If a *Seeker* wishes to experiment with the advanced version: start with the previous *process* as given while seated comfortably in a room. Once you have reached an *end-point* with that, close your eyes and repeat the *process* using an "imaginary" view of the room.

This practice is best done without straining and without concerns about accuracy. It is important, especially early on the *Pathway*, to acknowledge every "win" without invalidating a level of ability not yet regained. A *Seeker* should also avoid repeatedly opening their eyes to "check" whether or not their personal "copy" fits what is otherwise viewable with the body's eyes.

When first practicing with a technique like this, there are likely to be a lot of gaps of *real perception* filled-in with "created" or "ima-gined" scenery. Much of it will not necessarily be a one-to-one duplicate of what the body sees. It is also possible to perceive things that *are* "real," which the body is not able to sense.

Although presented early in our instructional lessons, this version is actually part of the advanced *processes* because its application is not restricted to standard *"defragmentation"* procedures. It is just one example of the *processing* we treat in this *Professional Course* that continues to be practiced at "advanced levels." It does not particularly have a "finite" *end-point* when applied during this lifetime.

Another application of this formula is to *knowingly* focus *attention* on specific points of the body (or *genetic-vehicle*). There is a tend-ency to operate the body on "auto-pilot." Often, our *attention* "snaps-in" on the body rather violently during painful incidents —and *then* we are suddenly *very aware* of it. The lack of *Self-de-terminism* involved in this abrupt shift in *Awareness* only reinforces the falsehood that we *are* our bodies.

HELPFUL TIPS ON PROCESSING

The second *process* we will introduce in the *Professional Course* is also an integral part of *Opening Procedures*. It is part of a cycle re-ferred to as "Command of the Mind-Body Connection." Where we previously have *knowingly Self-directed* our *attention* to "spot" things, we now make deliberately intended actions to both "reach" and "withdraw" from them. This will also allow for addi-tional training on general *processing*.

Perhaps one of the more challenging aspects for beginners to grasp about *processing*, is knowing just how long to *run* a *process* for. There is, of course, a liability to either *running* a *process* too long or not long enough.

For example: if a *process* is left as an incomplete cycle, a *Seeker* does not earn the gains or new *realizations* they otherwise would —and they likely have left a bit of *attention* on something stirred up, but not handled. On the other hand: if a *process* is run too long, a *Seeker* may start to feel tired (or "heavy")—the actual end-point when they felt better from the *process* was missed or unacknowledged.

There is much less liability—and by this, we mean the chance of hindrance of progress on the *Pathway*—by experimenting with "*overrun*" and "*underrun*" on these more fundamental "*objective processes.*" Getting a "sense" for this early on the *Pathway* allows greater certainty in handling more intensive *processing* further along. This is of tremendous importance for *Seekers* intending to *Solo-Pilot* the entire *Professional Course.*

This *process* immediately follows the previous one in the *Standard Opening Procedures* of a *Formal Session* because it builds on the *presence* and *certainty* already established. In order to "touch" and "let go" of an object, a *Seeker* must first "spot" the object in present space-time. This may seem like only a slight increase or gradual incline in the "challenge" or "difficulty level" presented to a *Seeker*—and rightfully so. Much of the stable progress earned along the *Pathway* will be attained this way.

The repetitive alternating "*processing command-lines*" ("PCL") given in the sample script for a *Formal Session* are:

"*(Choose an object.) Decide you are going to reach for it; then make that body pick it up.*"

"*Now decide when you are going to put it down and make that body put it back where it was.*"

Usually, once a specific object is chosen, the *process* is *run* on that same item repetitively. While this example is quite direct, there are many PCL variations that could be just as effectively applied. What's given above is not even the most basic form of this *process*, which is:

*"(Choose an object.) Decide on an exact spot on the object and reach
 for it."*

"Now touch that spot for a moment; and then let go of it."

This is repeated over and over quite a few times. In addition to its use as a preliminary to a *Formal Session*, typically a *process* such as this is *run* in order for a *Seeker* to actually learn something, or come to a "new" *realization*. This is always our intention for *processing*; but in this case, it is to *realize* the level of "command" an individual directly has over the behaviors of their *Body* and functions of the *Mind*.

Putting the *realizations* of this specific *process* aside for a moment, let us use it to demonstrate *running* a *process* in general. For one thing: a continuing *Seeker* (from the *Basic Course*) with an understanding of the "Beta-Awareness Scale" (detailing a sequential range of emotions and mental states) may notice that an individual often comes *up* from the *bottom* of the scale during a *process.*

It requires a bit of *Self-determination* to get started on a *process* and quite a bit more to continue *running* it. In this example, it takes some time *running* to get past the immediate feeling of "*So what?*" or "*I touch things all the time.*" After further *running*, perhaps the attitude rises on the scale to "*This is stupid*" or "*This is boring.*" But if you push through this, a *Seeker* gets interested and starts to feel better and the object seems brighter. **This** is when you have reached an *end-point* on the *process*.

"*Underrun*" would be anything short of the *end-point* (such as the other states or attitudes just described). However, if you want to get a sense of "*overrun*," you can experiment with this same *process* by continuing it longer, past the appropriate *end-point* when you were feeling good about it. The longer you continue to *run* it, the further back down the *Beta-Awareness Scale* you may find yourself slipping; feeling worse about it and finding the repetition less tolerable and more difficult to handle than before.

Such "*overrun*" can easily take place during *processing*, when an *end-point* is not acknowledged (and the *process* is not stopped). In *"Traditional Piloting,"* it is up to an experienced practitioner to recognize these points—but when *Flying-Solo*, a *Seeker* should get familiar with this phenomenon and know how to fix it.

The basic pattern observed of *overrun* is: a *process* begins "rough," then suddenly it "smooths" out and seems easy and fun (which is the *release-point* or *end-point*), and then it starts to feel "tough" again to continue doing. In this case, a *Pilot* has "flown past" or "bypassed" the appropriate "landing spot."

The most basic solution to *overrunning* a *process* is simply to "spot" the exact moment when you were feeling good and *did* reach an *end-point*. By "spot," we mean to definitively notice or perceive something distinctively—and not necessarily visually or with the eyes. If you can bring to mind the instance when you were feeling good about the object, then it should start to seem that way again.

At the other end of things, *"underrun"* is primarily a result of one's own premature *withdrawal* from the *process*. It only occurs by stopping a *process* before an *end-point* is reached. When not due to outside interruption, this usually happens when the content or data addressed by a *process* causes discomfort.

In the example we have been using, there is very little mental strain or emotional turbulence attached to touching and letting go of an object in our surroundings—unless perhaps we have selected an object that we don't particularly *"like"* very much. But in later *processes*, once something is critically stirred up, it is important not to *withdraw* from it simply because it seems "difficult."

The "difficulties" initially encountered with handling something "turbulent" are much different than how things seem when something is *overrun*. For one thing: if a *process* is *underrun*, a *release-point* or *end-point* has not yet been reached. So we don't want to *withdraw* from a *process* just because it feels uncomfortable.

Therefore, if something is "happening" (*e.g.,* there is movement on the *Awareness Scale*) or a *process* has "triggered" or "turned on" a reactive-mechanism, the only systematic action is to continue *running* the *process* until the *end-point* is actually arrived at. This is why we use "*sessions*" to avoid outside influence; because if the present *restimulation* is due to the *process*, then continuing to *run* that *process* will handle it. And, of course, the moment it is actually *handled*—when you "feel good" about it—you *end* the *process*.

TOUCH-AND-LET-GO

We use the "*touch-and-let-go*" exercise as a training example at the *Mardukite Academy*, but it is also a real *process*. The standard version has already been relayed:

"Look around and choose an object."

"Now, choose a specific spot on the object."

"Touch and let go of it (until you feel good about it)."

"Choose a different spot on the object; and do the same."

"Do this on individual spots until the object seems more acceptable to you."

After having some practice with this *process* on many "spots" using various different objects, you can then apply this technique in everyday life with various things you find yourself using frequently. This is especially useful on "things at work"—and also on automobiles; everyone should be required to do this on vehicles before they are driven among other individuals.

As a variation of this (in *processing*), it is quite standard to have a *Seeker* "spotting spots" on "walls" of the *processing* room, and performing the same "*touch-and-let-go*" cycles. This not only increases personal "*presence*" (of *Awareness*) in present space-time, but also increases the vibrancy of the room. Using "walls" in this *process* enables directing *Awareness* on "spots" that are not part of objects or particularly interesting.

In a *Formal Session*, "*objective processing*" (such as this) is also applied in between the more "introspective" *processes*, in order to maintain a *Seeker*'s orientation (or "*presence*") of *Awareness* (in present space-time). This is particularly important during prolonged *intensive sessions* that include many individual *processes*.

Advanced applications of this basic "*touch-and-let-go*" technique require "mentally reaching out" with *attention* — or extending/projecting one's *Awareness* — like a beam of energy.

As with handling the very first process given in this lesson, you would practice by beginning with the "physical" version (and *running* it to an *end-point*) and then switch over to the "mental" one (using the same object). This would first be *run* with eyes open: *looking* at the object and mentally reaching — not just with your *attention*, but as if you are actually making "contact" (or "touching") the object. It can also be *run* with eyes closed.

Another advanced variation of this is applied to the *Standard Opening Procedures* of a *Formal Session*, when a *Seeker* is directed to:

"Close your eyes. Put all of your attention on the upper two back corners of the room and just get real interested in them for a while."

WILLINGNESS TO REACH

One reason we practice deliberate (*Self-determined*) and repetitive "*touch-and-go*" cycles is because so much of this activity in our lives becomes "automated" and "reactive." For example: if we touch a "hot" surface, the reflex to pull away (retract or *withdraw*) is not *Self-determined*. This lack of *Self-determinism* even contributes to prolonging the sense of "pain" we may experience afterward.

One of the systematic techniques for handling this, which can be used in everyday life, is called the "*touch back*." By this, we mean duplicating the action that hurt you on your own *Self-determinism*.

Of course, we mean to do this *slowly* and *safely*. This should include repetitive *"touch-and-go"* on various objects in the location of the incident, in addition to whatever caused the actual injury. But, for example: in the case of a "hot stove," you would wait until it cooled; or you would cover the edge of a "sharp" instrument, *&tc.*

The automation or reactivity of a *"flinch"* or *"withdrawal"* also involves a decrease in *Actualized Awareness*. It is simply inherent in the systematic mechanisms at work. When you are hurt, *attention* may "snap-in" on the body as an *effect*, but the actual *Self-determined Awareness* withdraws. Even with emergency medical attention, the "withdrawal" or "avoidance" of *Self-determining attention* on that area will inhibit healing. Most physicians will agree that beyond the medical care they can provide, the rest is the patient's attitude.

For example: even with something as simple as "stubbing your toe"—you may have noticed that the sting of the pain continues to hurt after the incident takes place; and for a surprisingly long time, given the lack of real damage that usually occurs. What's more: there is a tendency to "stub" the same toe again soon afterward, because we have also *withdrawn* our attention from the toe and prefer not to *confront* the event.

The short answer to this "mystery"—and most aspects of *defragmentation* in general—is *"Awareness."* An application of the *"touch back"* in the above example would be to *Self-determine* moving your foot slowly to lightly contact the surface you had hit.

The first time, the sharpness of the *restimulation* brings up to mind the original incident itself. But, after a few more *runs* of this *process*, the pain more quickly subsides (and there is less of a tendency for the same injury to repeat again). This is one example of applying *Systemology* philosophy to life that does not require a PCL or *"session"* to utilize—*although* specific instructions may be easily communicated if assisting another.

What we are starting to handle, in the long run, is the *"willingness to reach"* —and the more "physical" or "objective" exercises systematically *duplicate* the type of *mental processing* already taking place in the Mind. There is a tremendous amount of theory behind these techniques that is covered in the more advanced *Systemology* texts; but the *processes* given in this *Professional Course* speak volumes for themselves if applied.

THE WALL

The final *process* covered in this lesson is deceptively simple to *run,* and is also popularly spoken of by *Mardukite Academy* students. There is always a lot of gossip and jokes concerning *"The Wall."* This sometimes leads to not taking it seriously; but that is usually a result of not executing the actions in a fully *Self-determined* and precise manner. Each of the actions is made as if it is the *first time;* not simply a *repeat.*

The most basic PCL for this *process:*

"Look at that wall."
"Walk over to that wall."
"Touch the wall."
"Turn around."

This *process* is *run* in an open room between two walls. It is important that there is a clear path between the walls. It is preferable if the walls are bare. It is done *over* and *over* again. There is a point when *running* this *process* can really *"ping"* you with some discomfort; it may even stir up a lot of sensations that stem from things that remain "beneath-the-surface" and not yet accessible. But push through any of this and continue. Whatever point you start to feel good about handling control of the *Body,* end the *process.*

As with other examples of *"objective processing,"* this may also be applied as an advanced "eyes-closed" *process.* In this case it is best if you can lie down comfortably on the floor. And again, as with the other examples, you would begin your practice with the standard version, physically walking between the two walls. Afterward, you perform the same deliberate PCL actions, but *imagining* yourself as a *Spirit* doing it.

LESSON TWO
"THOUGHT AND EMOTION"

KEEPING A FLIGHT LOG

Whether a *Seeker* is *Co-Piloted* or *Flying-Solo*, it is traditional to keep a *"Flight Log"* or *record* of the journey on the *Pathway*. This follows in line with the systematic approach of our philosophy. It is also helpful to refer to if running into any misunderstood *turbulence* along the way—such as *"underrun"* or *"overrun"* of a *process*, as described in the first lesson of this *Professional Course*.

A *Systemologist* keeps careful records of their *"processing sessions"* and any other *realizations* that have occurred while traveling the *Pathway*. In 2020, the *Systemology Society* developed the *"Truth Seeker's Adventure Journal"* to make this easier—but you can easily use any notebook to "log" your progress if you know what information is most critical to keep track of. Be consistent.

The following is a brief list of all general information tracked in *Traditional Piloting*:

–*Name* of the *Pilot* or *Co-Pilot*;
–*Location* and *Weather*;
–*Date* and *Day of the Week*;
–*Beta-Awareness* (at *start* and *end* of session);
–*Time* (at *start* and *end* of session);
–*Processes*, *Routines* or *Procedures* used;
–Any *Terminals* (*masses, objects, people, places*) contacted, recalled or imagined in session;
–Everything a *Seeker says* in session; and
–Any *Realizations* that occur.

Records are important for optimum effectiveness and tracing any incomplete *processes* or other sources of *turbulence* that may have become accessible, where they were not before. The techniques included in the *Professional Course* series are not meant to be particularly strenuous, however, even the most basic of these *processes* can "stir up" or "trigger" various things—each of which should be handled as they *resurface*.

Another purpose for *journaling* as a standard routine is to "*extrovert*" after a session; to get all the stuff that took place down on paper, externally. This also helps keep a *Seeker* from keeping any residual attention remaining on the session and not in present space-time once the session has ended.

CONTROL OF MIND AND BODY

Thoughts (of the "*Mind*") and subsequent *emotions* (of the "*Body*") are under the command of the *Alpha-Spirit*. Of course, we often find ourselves relinquishing the control of these aspects to others. Sometimes it is simply a part of the roles we play—or perhaps it helps to keep things interesting—but in the end, it is always within your power to regain total control of your *thoughts* and *emotions*. Some basic exercises in this area of "attitude control" are provided in the *Basic Course*, as is an introduction to the *Beta-Awareness Scale*. However, let us start off this lesson with a demonstration of light "*objective processing*."

"*Choose a 'neutral' object (that you have no strong feelings about).*"

"*Look at the object.*"

"*Decide to feel various positive things about the object (that you love it, that it is beautiful to look at, that it is wonderful to have close to you, &tc).*"

"*Maintain this positive attitude for several minutes.*"

"*(Then reverse this and) Decide to feel various negative things about the object (that you despise it, that it is ugly to look at, that it is harmful to be near, &tc) and continue this for several minutes.*"

"(Then reverse this again and) Decide to maintain a positive attitude about the object for several minutes."

When this *process* is practiced seriously, a *Seeker* may initially encounter some challenges in exercising their fluid alternation of attitudes back and forth. Various unintended "reactions" may also occur as one makes the decision to feel one way or another. This should be repeated several times on the same object until any automatic reaction or turbulence has "smoothed out"—or *"flattened"* as we tend to say (using *wave* terms).

Once this has been *run* and *flattened* on the same object, you may then use the *process* again on other objects in the room. Always end-off during the "positive" side of an alternating *process*, rather than the "negative." This *process* is not intended to solve all *fragmentation* preventing total control of the *Mind* and *Body*; but it is a very productive practice—and it may lead to the same *realizations* on which our theories for it are based.

ASSOCIATIVE IDENTIFICATION

Much of what *Humans* consider as *"knowledge"* really relates to *"associative identification"*—or more simply, *"association."* Entire fields of philosophy—such as *"epistemology"*—are dedicated to theories to explain *"how we know what we know."* *Systematic Processing* allows a *Seeker* a direct approach to determining the truth of these matters for themselves.

Early work by the *Mardukite Research Org* served as a precursor to establishing the *Systemology Society*—and within that body of early research, we uncovered a significant amount of information regarding the control of human consciousness with *"language"* at the inception of modern civilization thousands of years ago.

Most *"thinking"* involves *association*, not *Knowingness*. The type of *realizations* that occur in *processing* lead to a higher state of actual

Knowingness. This Knowingness is not the result of freewheeling thought—it is not a result of "thinking about" or "figuring on" things. The systematic approach we take in processing is similar to natural "thought processes"—but a Seeker more directly is applying attention to actually look at things.

The following process demonstrates all of these points. When first practicing this, it is most effective to select an object present around you that has a lot of "mass" to it—such as a table, a chair, or a bookshelf, &tc. Then, select something else with "mass" (not present in the environment) that it would be ridiculous to associate with the object—such as an apple, a garden-hose, or a stop-sign, &tc.

The basic "processing command-line" (or "PCL") for this exercise is:

"Look at the object and immediately think of the ridiculous association you have chosen."

  –or–

"Look at the object and think of the ridiculous association as if it were automatic."

To be most effective, it is best if you alternate doing this with looking at something else around you and noticing something about it. Then you can redirect your attention back to the object you are working with, and getting an immediate sense that it is the ridiculous alternative association instead.

As an advanced application, you might try to actually imagine or visualize the intended alternative in place of the object. When you "spot" this imagined alternative, make a point to notice specific things about it. This will also increase the detail or vibrancy of your visualization. This may be practiced further using various different objects in your environment and associating other alternative ridiculous items.

Usually we associate "ideas" and "concepts" in our "thinkingness," but the actual substance or significance we attach to our thoughts and feelings often regards something with "mass"—which is to say, a "*terminal*" using *Systemology* vocabulary. We treat specific *terminals* more directly in later "*subjective processes.*"

CREATING EMOTIONAL FLOWS

Strength and ability of an *Alpha-Spirit* is rooted in *Self-determinism* —of being *at cause*. Therefore, we find that misappropriating (or inappropriately assigning) *cause* in our life is one of the many ways in which we increasingly come to consider ourselves primarily the *effect* of others, and even the *Universe* as a whole.

It is quite customary with the *Human Condition* to assign *cause* of "ill effects" as far away from *Self* as possible. From this, we tend to "blame" others for how we think and feel. This is not to say there are not many influential factors to our experiences; but ultimately, it is *Self* that decides how to think and feel.

Whenever we "incite" or "inspire" an *emotional reaction* in another, they sometimes tell us that we *caused* them to *feel* this or that way —that *they* are experiencing the *effect* of our *presence*. Of course, many factors are likely at play here; but the ability to *create* a specific *emotion*, *sensation* or state of *thought* is certainly within the power of an *Alpha-Spirit*.

With *objective processing*, a *Seeker* can *knowingly* practice getting this sense of *creating* or *projecting* an "*emotional flow.*" This will also start to familiarize a *Seeker* with the idea of more directly considering (or treating) all *energy*, *communication* and *attention-lines* as a "*flow.*"

"*Look around and choose an object.*"

"*(Alternate) Make it feel happy; Make it feel sad.*"

When practicing this, perform the action by *intention*. You don't need to actually "say" anything—telling jokes or scolding—to produce the effect. Just get the sense of deciding upon, and intending, the idea or concept very strongly and clearly. Then, get the sense that just being in your presence makes it happen.

After this is practiced for a while on a specific object, try it on other items in the room. Once you are comfortable with this pair of states, consider other states on the *Beta-Awareness Scale* such as *"love"* and *"hate"*; then with *"interest"* and *"boredom."* Starting with the extremes will eventually allow you to easily practice with any emotion or mental state.

When we interact with the *Universe*—and other lifeforms in the *Universe*—*communications "flow"* in various directions between *terminals*. In *Systemology*, we refer to these individual *flows* as *"circuits"*—therefore *"circuits"* between *"terminals."* We adopted this terminology from the field of electronics; but, this systematic approach has allowed us to understand and apply our philosophy in *systematic processing*.

"Look around and choose an object."

"(Alternate) Get a sense of feeling sorry for the object; get a sense of the object feeling sorry for you."

Practice for a while on the same object, then pick another item and do the same. If a *Seeker* continues this long enough, the ultimate *end-point* would be a greater control over the *reactivity* associated with *"grief"* and *"sympathy."* A complete mastery is not expected with a single pass through the *Professional Course*.

This *process* may be applied to other states of *Beta-Awareness* —"fear" or *"being afraid of"*; "anger" or *"being angry at"*; "cheerfulness" or *"happy to have you/it there"*; *&tc.* Again, working with these states *knowingly* in *processing* should produce some sense of having a greater control over the experience of them.

AUTOMATION AND REACTIVITY

One of the principle areas of *Systemology* is the study of *automated mechanisms* and *reactivity*. In basic terms: this is how *systems* influence one another. Most systems are *dynamic*, which means that they do not exist in exclusion separately from other *systems*; thus their *"state"* may be influenced by other outside *"conditions."*

The *Alpha-Spirit* has the ability to command the control of all its *systems*—and the ability to automate that control as well. While the *"Body"* does receive its commands from various Mind-Systems, much of what is handled by the *"Mind"* takes place on *automatic*, leading *unknowingly* to act out behaviors as a *reaction*.

A systematic way of handling those tendencies of the *Human Condition* that are happening on *automatic* is to *knowingly* "exaggerate" their behavior. This is far easier than invalidating early efforts to "stop" it altogether at first. Then, in practice, you alternate doing it more and doing it less. You basically keep doing the thing *knowingly* until it seems to come more under your control.

There are many aspects of the *Human Condition* that are experienced *"on automatic,"* but this basic technique is quite effective for handling the most accessible (visible or already known) tendencies. Let's try another exercise within the same area of *processing*.

"Close your eyes."

"What are you looking at?"

Some individuals see a specific *mental image* (*visual picture*), or *energy-bursts*, or even *blackness* (but a uniquely intense *blackness*) when they close their eyes. This occurs seemingly automatically. The *process* continues:

"Make a duplicate copy of what you see (right next to it)."

However a *Seeker* manages this action is acceptable. It might be an *imagined* "holographic" duplicate; or perhaps you imagined a "screen" (as a background) to place a copy on. At first, we are really only concerned with getting a sense of this, even if the *Seeker* has difficulty with creating vivid *mental images*.

Once a single copy of whatever is seen is made next to it, make another duplicate copy on the other side. Now make more copies to the right and to the left. Don't limit the space to a linear horizon in front of you; make more copies above your head and below your feet and behind you. Then continue to the next *process*.

"Close your eyes."

"What are you looking at?"

"Make a mental image copy of it."

"Now change its color; Turn it ---."

Whatever the color-quality already (or *automatically*) is, "*turn it...*" to another color. In between cycles of this *process*, be sure to open your eyes and look around your environment, spot things and notice specific points on them. This alternates using purely "mental" exercises with reorienting *presence* in that space and time. As an advanced application, a *Seeker* creates many copies and starts changing the color of each one individually.

We have mentioned "*association*" previously in this lesson; a large amount of our *associated knowledge* is made up of "automatic" and "reactive" *mental imagery* that represents the "terminal" or "mass" that we *most* "*associate*" with that concept or piece of data. Often, this restricts us from considering a wider scope (or range) of potential possibilities.

There is another phenomenon observed in *Systemology*, referred to as a "*compulsively created mental image.*" This is otherwise called a "*stuck picture*"—one that is consistently brought to mind, or *unknowingly created*, outside of, or without, one's own *Self-determinism*. Most individuals have a few of these that are *imprints* of incidents and past experiences. More often then not, a "*stuck picture*" is of something unpleasant.

For example: if you are worried about something, there is often a *"mental image"* representing whatever you are worried about that is *"stuck in your mind"* —which is to say, that you can't stop thinking about. The systematic solution is to make many copies of the *"stuck picture"* —first changing its color, then altering the entire nature of the imagery one part at a time until the whole "scenery" comes under your control.

"Look around (the room) and spot an object."

"What thought comes to mind?"

"Make a duplicate copy of the thought."

"Make another copy (make many copies, thinking it many times)."

"(Alternate) Make the thought 'louder'; Make the thought 'quieter'."

Having the nature and intensity of thoughts under your control will assist in better handling those aspects and items we'd rather not *confront*.

CONFRONTING "AS-IT-IS"

The word *"confront"* has many negative connotations in modern language. For example, we tend to refer to someone who is quick to exhibit anger with others as *confrontational*. But this only partly demonstrates the true meaning of the word, which is: "to come around to the front" —or "come around to face."

In some ways, we have been lightly (but directly) *"confronting"* objects, emotions, and states of mind throughout this lesson. We have demonstrated that control over our experience of something, lies not in our avoidance or withdrawal, but our willingness to *look* at it *"As-It-Is."*

The first basic *confront processes* (also known as *"concentering"* in archaic *"New Thought"* traditions) are important, but easily dismissed or incorrectly performed. The ability to properly *confront* existence *"As-It-Is"*—and free of automatic-associations and other reactivity—is a critical component to the *Pathway*. But, this does not take place all at once, just as the "blockage" to this native ability did not always seem as solid.

When we *look* at something *"As-It-Is,"* we are *confronting* it. The ability to do this in *Self-Honesty*—without those *associations* and *reactivity*—is a practiced skill, one that improves the further along the *Pathway* one progresses.

Not unlike states sought through traditional forms of *"meditation,"* to *confront*, we practice by simply *looking* at something with all of our *presence* and *attention* without any *distraction*—either from the *"Body,"* the environment, or thoughts. This also increases personal ability to provide *presence* and focus *attention*.

There is an archaic *Tibetan* tradition of *"Sitting Face-to-Face"* which a *Seeker* may practice best with a partner. This is performed by simply sitting across from another person without talking or giving in to reactions. In *session,* this may be practiced with an object.

What is important is to be able to "extrovert" your *attention*; to project it to a point that is outside and away from where you *perceive* the *"Body"* or *"Mind"* to be located.

Unlike the previous *process*—where we "make a copy" of distractions (thoughts and emotions) to practice getting them under our control—this exercise is not concerned with addressing each distraction as a *thing*, but rather ignoring them and refocusing *attention* exactly where you want it. You might be surprised just how much "stirs up" in the *"Body"* and *"Mind"* just sitting quietly maintaining eye contact with someone.

If any discomfort, emotional reactions or intruding thoughts "stir up" or "turn on" while performing this *process,* simply maintain focus of your *attention* "outwards" until they "simmer down" or "turn off." There are many times in *Life* when you need to persist or handle something in spite of distractions, and the ability to do this *clearly* without losing *focus*, is an important skill for optimum *survival* (and our ability to *create*).

This may be practiced until you are able to achieve these results effortlessly for a few minutes at will. Afterward, reapplying the *"Control of Mind-Body"* process used in the *"Opening Procedures"* of a *Formal Session* (and reintroduced in *Lesson-1* of this series) may yield different results. The PCL for this is:

"Put all of your attention on the upper back corners of the room; Keep your attention there and ignore all distractions."

INTRODUCING SUBJECTIVE PROCESSING

The sequential *Pathway to Ascension* relayed in this *Professional Course* series of lessons begins with a lot of *"objective universe processing."* This means those *processes* which mainly "extrovert" *attention*, or focus it *"external"* to the *"Body"* on the *"objective"* Universe. Even when treating something *"internal"* (like emotion and thought), we have, up until now, *processed* a *"projection"* of it into *"external"* objects.

The majority of *processing* used on the *Pathway* concerns the *"subjective universe"* of the individual themselves. This might be best understood as a *Seeker's* own *"Personal Universe."* However, all of the previous *processes* allow for early personal development, are excellent demonstrations of *Systemology,* and quite useful for re-aligning full *presence* in between long periods of "introverted" *subjective processing.*

Conceptual Processing is a form of *subjective-universe processing* because it asks a *Seeker* to "consider," "conceive," or "get the sense of" a concept being (or not being) a certain way, in spite of how it might really appear or manifest in this *objective-universe* (that is otherwise agreed to as *"Reality"* for all concerned).

Since *subjective-universe processing* is such an integral part of the remaining *Professional Course* series, it is important that we introduce it, and its systematic application, in the earliest lessons. There is also some additional instruction that is necessary, which a *Seeker* should combine with what they've already learned about *systematic processing* in general. This is important for *Flying-Solo* without experienced assistance.

Systematic Processing is *not* a form of meditation or psychotherapy; and it most certainly *isn't* freewheeling thought, where an individual just skips from one thought to another or lingers compulsively on one. Our methods originally developed from a need within the *Mardukite Research Organization* to provide *Seekers* with an effective form of ministerial counseling or spiritual advisement. *Systemology* developed thereafter.

When working *Solo*, it does not help to start "spinning in" on endless cycles of *"thinkingness"* that ultimately leave you more confused, or at the very least, less certain, of your ability and your progress on the *Pathway*. It is *certainty*, and the brighter "feel good" *objective processes*, that will help carry you through the rough spots.

To avoid stimulating anything significant, we will introduce *subjective processing* with a "neutral" example that would not otherwise appear in a *session* and should not have any mental *"charge"* or emotional turbulence attached to it.

A *Pilot* gives (or a *Seeker* reads) the PCL: *"Think of a Fruit."* Now, the *Seeker* would either communicate (or write down) all the *"answers"* that come to mind from this. The two-way communication (or writing, if *Solo*) assists in keeping a *Seeker* from getting too introverted or "spun-in" on a *subjective process*.

At first, in this example, a *Seeker* might have to orient themselves, or internally *process* the *command-line* for a moment, and then the initial *"answer"* (or *"response"* to the mental inquiry) might be "a cherry." And so, this would be communicated (or written) the moment it comes into mind. This also allows that cycle to get acknowledged, and the *Seeker* is free to come up with another *answer*. Soon, many of them start to occur quite easily. The *endpoint* here is when you feel good about *"thinking of fruits."*

If this were a *process* regarding something with *fragmentation* attached to it, there would initially be some noticeable difficulty or resistance in *running* the PCL—or the *"answers"* wouldn't make sense, or would be about something else. *Systematic Processing* increases a *Seekers "certainty"* in being able to push through the "mental barriers" and regain control of "emotional turbulence."

An example of *fragmentation* if *processing* our example, would include "spinning-in" on free-thought that concerns the nature, or definition, of what a *"fruit"* is, *&tc.* This is one of the reasons an experienced *Pilot* will make certain that a *Seeker* fully understands the words used in a PCL before using it in *processing*.

Free-thought, or freewheeling thought, is what takes place due to *associations* and unfocused *attention*. For example: a *Seeker* answers (or writes) "a cherry" and then starts (internally) asking themselves questions about "cherry-pies," which leads to wondering if they are hungry, and so on. This is not *processing*. This does nothing to *flatten* the *"wave," "charge"* (or *"turbulence"*) of the *fragmentation* presenting itself.

When *running* a *process*, you want to focus on getting direct *responses* or *answers* to the PCL; even if it's not posed as a "question" outright. This is what makes our work a *systematic process*; because you are *looking* at *responses* or *answers* from a "computational databank." These are not random *responses*, but *answers* that result from *systematic processing* taking place "internally" in the *"Mind"* due to *running* our intended PCL. During a *process*, the PCL is the query or question, which already puts a *Seeker* in some degree of *uncertainty*. The key to *running* a *process* is to then produce *responses* or *answers*, not pose more questions to yourself. If you consider each PCL in a *"What is..?"* form, you should be most focused on *"It is..."* answers.

BASIC SUBJECTIVE PROCESSING

To further illustrate the application and proper *running* of *subjective processing*, we will begin with a few light examples that are real
processes. Alternating *command-lines* may be designated "A" and
"B" (or however many steps there are to the *process*. These can be
logged in a journal simply by putting the letters at the top of the
page and using each subsequent line to keep track of the appropriate list. Therefore:

A. *"Think of something you wouldn't mind remembering. (What is it?)"*
B. *"Think of something you wouldn't mind forgetting. (What is it?)"*

This is systematically more effective than simply drilling on one
consideration repeatedly. A *process* is already *run* in repetition, but
doing this with *subjective processing* also allows you to realign your
focus and avoid getting "spun-in" or "off-track" on some additional cycle of thought.

There are also *subjective processes* where you do simply keep listing *"answers"* from one PCL until one seems right—but that is
only used when you are scouting for a specific *"answer"* and not
unraveling a whole chain of *considerations*. This type of work will
be treated later on in the *Professional Course* series.

In *running* our present example, you may find that answers do not
immediately present themselves, but eventually they do—and at
one point, perhaps quite quickly. There will then be a point when
new answers don't seem to be occurring and suddenly the *process*
seems difficult. The *realization* or improvement will usually come
by pushing through this at least once, *looking* a bit "deeper" to see
if there is something else.

"What is something you agree with?"
"What is something you disagree with?"

Using *subjective processing* for *defragmentation* generally consists of "freeing up" *considerations* that may otherwise be rigidly fixed in place as they are. This rigidity or solidity produces what we may refer to as a *"stuck flow"*—which is fixed continuously in one direction. The systematic solution to this is to practice an *alternation* of *consideration* from multiple sides or angles. Without understanding this, some of our *processing* would be seen as counter-intuitive. For example:

"Decide that something is important."

"Decide that something is unimportant."

In *running* this, we aren't as much concerned with what that "something" is for a single *process*, or what the *Seeker* ultimately feels about it. The entire purpose is to practice *considering* that something *is* or *isn't*. As a result, an individual is more certain in their ability to actually "change their minds" of their own choosing.

For the final example in this section, separate your journal page into four columns—A, B, C, and D—and just think about something related to the categories given for each PCL. For each PCL, you will think about something and then list what *"it is"* as if responding to the follow up query of *"What is it?"* The answers are not as important as the practice of maintaining *Self-determined* control over your thoughts.

A. *"Think about Space."*

B. *"Think about Time."*

C. *"Think about Energy."*

D. *"Think about Matter."*

CONFRONTING THE PAST

Most of the *processing* presented in the *Professional Course* series will target specific areas. This is important for learning the various *processes* and *techniques*, and also for developing *certainty* and increased *Awareness* for when a *Seeker* cycles through the material of the entire course additional times.

Stable progress on the *Pathway*—and *processing sessions* in general —results from being able to handle (or *process*) whatever presents itself. A *Seeker* learns to *confront* rather than *withdraw* from their own *Mind*. But before that can happen, it is necessary to have *certainty* on the right *processes* to apply to the right situation.

Fragmentation is an accumulation of debris or blockage that occurs gradually over time. As our spiritual philosophy relays, the individual (or *Seeker*) as an *Alpha-Spirit* has been around for a very long time. It has followed its own path of existence as an *Awareness* that we refer to as a *"Spiritual Timeline."*

The *Spiritual Timeline* extends like a "track"—and it includes all of our *"past"* existences (as an *Awareness*) far and beyond *this* physical incarnation (lifetime), or even *this* Universe. We presume that it extends well afterward into what we would consider the *future* as well. The part which is *"past,"* we call the *"Backtrack."*

We have mentioned *"stuck pictures"* and *"stuck flows"* or other rigidly fixed forms of *fragmentation* that limit our total potential and range of *considerations*. These actually accumulate as "energetic-masses" on our *Backtrack*. We often refer to these masses as *"imprints"* when they are *associated* with a specific *incident* or *terminal*. The more *imprints* accumulate, the more *fragmented* the experience of *Life* and *Existence* becomes.

Eventually, to fully *Self-Actualize* for *Ascension*, an individual needs to *clear* the channels of *fragmentation*, particularly those that include *"past trauma."* If *Flying-Solo*, the best course of action is not to simply *fly* headlong straight into the trauma, but instead, to first build up *certainty* and *inertia* (or else, personal *horsepower*) with the basic *processes*—and *recalling* pleasant times to balance-out handling of the unpleasant.

To conceptualize the *Backtrack*, think of a long reel of "movie film" that is systematically categorized with various dates and incidents —along with the *considerations* that we made as a result of those events. An individual (*Alpha-Spirit*) is as well off as they have full *Awareness* on the contents of their own *Backtrack*.

Those energetic-masses that accumulate actually block our clear view of the *Spiritual Timeline* or *Backtrack* and inhibit us from experiencing and handling better recall of the data. Although *attention units* of our *Awareness* are entangled up in the compulsive unknowing creation of these *imprints*, our resistance to wanting to *look* at them eventually turns into automatic tendencies that provide us only "*blackness*" to see. Obviously, our existence has not only consisted of unpleasant experiences—and it is on these that we should put our attention in order to reduce any "*blackness*" that is associated with our history, in this lifetime or otherwise. When *Flying-Solo* without assistance of an experienced *Pilot* (or *Co-Pilot*), use *recall* of "pleasant memories" if handling past trauma suddenly becomes too overwhelming.

There is the occasional phenomenon where a PCL directed at pleasure moments will instead trigger associated thoughts of loss, &tc. Although a *Seeker* doesn't want to get into a habit of *withdrawing* from such *reactivity* without handling it, the *process* only works by completing the PCL, and not diverting *attention* each time a *distraction* arises. We have practiced this already. If it happens, simply acknowledge that it exists and take it up in a later *process* or *session*.

ANALYTICAL RECALL

"*Analytical Recall*" processing is one of the first methods developed for applying our philosophy. At first, a *Seeker* needs only to *recall* or "remember" some aspect that the PCL calls for and then notice something about it.

As a *Seeker* becomes more experienced with *subjective processing* they will be in the habit of spotting various *"facets"* of a memory with these *processes*. These *facets* might include: time of day; location; living things present; emotions felt; any data sensed, such as smells, the quality of light, humidity—the list goes on.

There are four *processes* below. Each *process* consists of three PCL. These are *run* repetitively as an alternating cycle (1, 2, 3, 1, 2, 3...) until you feel good about doing the *process*. Then end the *session*, or go to the next *process*. When first practicing, focus on just spotting specific *"times"* and *"people"* (and any *terminals* of significant *"mass,"* such as a building, *&tc*).

Communication:

1. *"Remember a time when you enjoyed talking to someone."*

2. *"Remember a time when someone enjoyed talking to you."*

3. *"Remember a time when you saw two people enjoy talking to each other."*

Agreement:

1. *"Remember a time when you agreed with someone."*

2. *"Remember a time when someone agreed with you."*

3. *"Remember a time when you saw two people agree with each other."*

Liking:

1. *"Remember a time when you liked someone."*

2. *"Remember a time when someone liked you."*

3. *"Remember a time when you saw two people like each other."*

Understanding:

1. *"Remember a time when you felt that you really understood someone."*

2. *"Remember a time when you felt that someone really understood you."*

3. *"Remember a time when you felt two other people really understood each other."*

LESSON THREE
"CLEAR COMMUNICATION"

WILLINGNESS TO REACH FURTHER

Stable progress on the *Pathway-to-Ascension* is marked by states of increased *"Knowingness."* By this, we mean what a person *actually knows*. This *Knowingness* is quite different from what we are *told*, or other *associative knowledge*. We mean specifically: what a person already *knows* about *Self*, their *past*, *Life*, and all *Existence*—but, for whatever reasons, has *"blocked out"* from their present *Awareness*.

The long-run of the *Pathway-to-Ascension* is intended to return to an individual the *certainty* and *Knowingness* of their original native *"god-like"* state as an immortal *Alpha-Spirit*. We do not expect this to happen all at once; and there are many safe-guards of the Mind-System that prevent the flood-gates of total *Knowingness* from overwhelmingly *"caving-in"* on the *Seeker* all at once.

Spiritual fragmentation (which also includes matters of emotion and thought) accumulates beneath the surface of what an individual is presently *aware* of. Some use the words "unconscious" or "subconscious"—but these are not truly *systematic* terms. Yet, we do mean what is happening *unknowingly*. In *Lesson-2* of this *Professional Course* series, we introduced the idea of a continuous *"Spiritual Timeline"* that the individual carries with them as a "memory" of their eternal existence as an *Alpha-Spirit*. This includes the experience of *this* incarnation or "lifetime" as well as all others.

Having the entire memory of one's past—or *Backtrack*—*"resurface"* on them in one flash instant would be too overwhelming to behold. But it can safely occur gradually—and *Systematic Processing* is intended to help gradually restore *Knowingness* of the full basic state or *identity* of the individual as a *Spiritual Being*.

Of course, to accomplish these goals, one of the first requirements is that a *Seeker* actually be *"Willing to Know."*

In the previous *"Fundamentals of Systemology" Basic Course* series, we described the totality of *Awareness* as a "spectrum" divided into two main areas: what is clearly *known* and a dark area that is *not-known*. The dividing line between forms the basis for what is considered *"above"* or *"below"* the *"surface"* of *thought.*

This "line" is really a philosophical construct; so we aren't trying to move the "line." In *processing,* however, we are working to shift more of the *data* from the area of *"not-known"* to the area of *"known."* There is also a "gray area" of what is *almost-known;* what is *accessible* to a *Seeker* but remains *just* "beneath-the-surface."

The greater the *Willingness-to-Know,* and the more that is within an individual's *tolerance* to reach for, the "wider" or "larger" this *gray area* will be for what is *accessible* in *processing. Systematic Processing* both *accesses* what is *accessible* and increases the individual's *tolerance* to *confront* addition layers beneath it.

Fragmentation is handled as a series of layers—each layer representing a level of blockage. We don't usually make stable gains by simply digging a deep hole to what is buried far beneath. Too much of what surrounds at each layer will "cave-in" on the *Seeker.* Therefore, we strip away the debris in layers to expose the entire area underneath.

We have introduced the fundamentals of *systematic processing*—specifically *subjective-universe processing*—in the previous lessons of this *Professional Course.* Included with this is advice and tips for handling *processes* as a *Solo-Pilot,* and a few maneuvers for getting yourself out of trouble if you encounter turbulence. With that in mind, let's start this lesson off with some light *processing.*

[Note: if you have already attained the ultimate *realizations* as *endpoints* for any particular area (from a previous pass through this course material), then your practical instruction is to *"spot"* the moment it happened.]

EXPANDING WILLINGNESS

For *processes* like these, you want to *run* through as many cycles of the *"processing command-lines"* ("PCL") as you can, rather than dwelling on each individual answer. You want to generate a response as if it were an item on a list, then go to the next, rather than free-wheeling into an entire narrative or explanation.

Each *process* that follows here will consist of three PCL. These are *run* repetitively as an alternating cycle (1, 2, 3, 1, 2, 3...) until you feel good about doing the *process* (which is the *end-point* of the process).

The ultimate goal is to genuinely increase one's *Willingness-to-Reach* in whatever area is being treated, since ultimately a "*god-like*" being would be *willing* to *know, do, communicate* or *experience* anything, whether or not they actually choose to. There is another aspect to these *processes* that increases *willingness* to "grant" or "permit" others their own "*Beingness*" — to also allow others the freedom to *Be.*

Willingness To Find Out

1. *"What would you be willing to find out about yourself?"*

2. *"What would you be willing to find out about someone else?"*

3. *"What would you be willing for someone else to find out?"*

In this *process*, the phrase *"find out"* could be substituted with *"know."* And *"would you be"* is a basic, less intrusive, wording that is often used with beginning *Seekers*; but a more direct PCL approach is *"are you."* We will apply the direct approach here:

Willingness To Have

1. *"What are you willing to have?"*

2. *"What are you willing for someone else to have?"*

3. *"What are you willing for others to have?"*

In this instance, *"someone else"* means a specific person (*terminal*); whereas *"others"* is meant to include everyone in a particular group, for example: all other *"Humans."* Let's do some more of this *processing*.

Willingness To Do

1. *"What are you willing to do?"*

2. *"What are you willing for someone else to do?"*

3. *"What are you willing for others to do?"*

Willingness To Be

1. *"What are you willing to be?"*

2. *"What are you willing for someone else to be?"*

3. *"What are you willing for others to be?"*

When we speak of *willingness* and *accessibility*, we tend to also use the word *"tolerance."* This means what is within an individual's *willingness* to *confront*. One of the areas that a *Seeker* may actually *process* for greater *tolerance* in general regards "change." The most basic *process* is to *alternate* the following PCL repeatedly.

A. *"What would you be willing to have change?"*

B. *"What would you be willing to have remain the same?"*

An *objective* example of this same *process* is:

"Look around the room; Spot some things you would be willing to have change?"

"Spot some things you would be willing to remain the same?"

One of the reasons that so much of our existence remains in the realm of "not-known" is because of our avoidance of actually *confronting* the contents of "what lies beneath." In many cases, there is a deeply laden *fear* that inhibits our *Willingness-to-Know*. Or a person stops *"looking"* and starts "thinking" and "worrying" from a point of confusion instead.

Knowingness is preferred to *thinking*. There's nothing inherently wrong with *thinking*, except that it usually originates from a *fragmented* state and is used to substitute actual *Knowingness*. The worst fears generally concern what is "*not-known*" (or "*unknown*"), and not what an individual actually understands or *knows*. So, let's just get some of that out in the open with this next *process*, before we continue.

"Think of, or imagine, a horrible 'truth' that you might find out."
"What would be the consequence of that?"

SYSTEMOLOGY AND COMMUNICATION

Communication is a central subject to *Systemology*, because we consider *all interactions* between "systems" to be a "communication." In fact, "*Systematic Processing*" is entirely based on our understanding of communication from within the philosophy of *Systemology*. And in many ways, we are really treating various aspects of communication consistently all along the *Pathway*.

In *Systemology*, "communication" is *systematically* handled as a *flow*. This allows us to treat all types of communication; not simply the "speech" and "gestures" we quickly associate with it, in terms of the *Human* experience. *Flows* occur on a "*channel*" between two *terminals*—usually *you* and something else.

A *Seeker* may notice similarities between our *Systemology of Communication* and the way in which "*water motion*" or "*electricity*" is understood in other applications. Either of these could be used to demonstrate our principles. We will focus on "water" for the moment, rather than assume an understanding of *electricity*.

Running water is a type of *communication*. Although we treat it all inclusively as a "body" of water, it is actually composed of individual droplets, each of which might be thought of as a single "unit" of water. In a stream, the activity of *flow-motion* represents the communication. It allows a single "unit" to cross a *distance*, from a source-point to a destination-point within a certain period of *time*. Therefore, *motion* and *time* are connected.

Fragmentation is that which "blocks" *free-flow* on a *channel*. It creates a "dam" for an otherwise fluid current. And as we know, concerning water blockages and dams, this has a tendency to build up "pressure." In our *processing*, this "pressure" is equivalent to the "*charge*" or "*turbulence*" that is encountered for a given area.

Inhibited communication, and control of these *flows* by outside (*other-determined*) sources, is what leads to our greatest upsets in life; it leads to violent protests, compulsions, reactivity, automation, and other societal misfortune.

Fortunately, control over these "*communication barriers*" ultimately remains our own. For, regardless of the reasons, it is ourselves (as an *Alpha-Spirit*) that decides to go "out-of-communication" with a *terminal*, or to relinquish *knowing* control over the *flow* on a particular *channel*.

With enough *actualized* or focused *attention*, you can systematically push through any personal *communication barrier* without even having to address all the reasons for its being there. In the case of "basic" *Human* communication, most barriers are crossed simply by *quantity* and *volume*.

For example: if you were to talk enough about a particular area, you would eventually find yourself uninhibited about communicating that subject. The reverse of this, or how we become inhibited, is when an "outside" (*other-determined*) source repeatedly demands (or even enforces with actions) that we *stop* that *flow* of communication each time we *start* it.

Systematic Processing resolves this by encouraging a *flow* until the barriers are cleared away. This is one of the benefits to *Traditional Piloting* for some *processes*; because another individual is there to keep those cycles of communication *flowing* along a *channel*, without the *Seeker* feeling the need to *"hold back."*

Outside of our philosophy, a more familiar *"New Thought"* technique involves a practice of writing letters to individuals that we have difficulties communicating with—or about certain subjects. We don't actually send these letters, but it gets us to externalize a flow (out on paper) that is otherwise being held in. There is no pressure to actually engage in communication with the other person until we feel comfortable doing so.

If we were to extend this practice *systematically*, we would also include writing letters from the perspective (or *point-of-view*) of the *other* person, as if they are writing *to* us. This allows us to handle both the *"out-flow"* and the *"in-flow"* on a *channel*, between us and a *terminal*.

In *Systemology*, we also refer to these various *flow*-types as the *"circuits"* of a *channel*. These circuits are usually numbered: 1, 2 and 3. They correspond with the numbers given to the three PCL in many *processes*. They concern:

1, *out-flow*, what we project or send;

2, *in-flow*, what we receive; and

3, *cross-flow*, what we perceive or observe of others.

Communication is experienced on these three *circuits* of a *channel*; and on these same *circuits* we store our *fragmented data*. Let us see this directly in an example of *"communication processing."*

1. *"What would you be willing to say to someone?"*

2. *"What would you be willing to have someone say to you?"*

3. *"What would you be willing to have someone say to others?"*

The PCL are *run* in rotation. The procedural instruction is to *"spot"* specific things that you are willing to communicate about. It is possible that there are many specific areas that require additional *processing* later in order to handle completely, so be sure not to in validate the progress and gains that you actually do make along the way by feeling like you are still "avoiding" certain things.

This *process* is *run* until you feel an increased freedom in ability to communicate. The ultimate *end-point* on this would be total uninhibited communication about anything—regardless of what you actually choose to, or not to, communicate about.

An advanced upper-level application of this same *process* is included for consideration.

1. *"What would you be willing to 'read' in someone's mind?"*
2. *"What would you be willing to have someone 'read' in your mind?"*
3. *"What would you be willing to have someone 'read' in another's mind?"*

COMMUNICATION PROCESSES

The subject of *communication* is handled in various areas throughout the entire *Pathway*. Elsewhere, in more advanced material, we also treat movement of a *"particle"* as communication. For now, let us focus on what is most familiar for the *Human Condition*; mainly, observable communications that originate from a *lifeform*.

What is considered *magic* or *mysticism* is simply a handling of *communication* originating from an *Alpha* "spiritual" existence. This includes our own *"Alpha Thought"* or *"postulates"* as a *Spiritual Being* (separate and superior to a constructed *"Mind-System."* A *postulate* is a "decision for things "to be" or "not be." It does not originate from a "brain" or anywhere within *this* Physical Universe; it *impinges* on, or *perturbs* activity in, this Universe via a *communication* of *intention*.

Communication is a broadcast, projection or *out-flow* of something from one point to another, across some distance. In *Human communication*, the "words" and method of delivery are secondary factors to the actual *intention* itself. Before any words are chosen, or any visible activity occurs, an *intention* is made. There is also an intended *receipt-point* or "destination" for the *communication* to "arrive" at.

For example: when these *Professional Course* lessons are developed, there is an intended "message" and an intended "audience" for it. The choice of words and arrangement into lessons then follows thereafter. A *true communication* cycle only occurs when its "meaning" is *duplicated* at the receipt-point exactly as intended. *Professional Pilots* must be expertly trained and skilled in this area to deliver *processing* to others.

Where *communication* concerns the individual or *Seeker*, we are most concerned with *intention*—the increase of strength and clarity of *intention*. With enough *intention* behind the meaning, you might even say the wrong words and others will still be able to *duplicate* the understanding implied. There are all various kinds of phenomenon in the area of *communication* that we will practice here.

"Choose an object."

"Say 'Hello' to it repeatedly."

"Notice the point in space you are projecting each 'Hello'."

"Directly intend them to various specific points surrounding (or next to) the object."

"Now intently focus them right into the center of the object."

This is practiced with different objects until a *Seeker* feels they have a handle on projecting *intention* into specific points. Once this is practiced with a deliberate concentration of focus, a *Seeker* then performs the action more rapidly with various objects throughout the room, and *intending* to land the *"Hello"* in the direct center of each without having to strain (or lingering for more than a moment on one point).

"Invent a nonsense word and intend for it to mean 'Hello'."

Practice the previous exercises, but speak the "nonsense word" as you intend your *communication*. Then practice this with other random words that you intend to mean *"Hello."* The *end-point* on this is a greater sense that *meaning* and *intention* is separate from the words and sounds *communicated* from the *"Body."*

Practice projecting *intention* with force as you shout *"Hello!"* at the objects. Alternate this with "whispering" it; but the emphasis here should be on sensing the same strong *intention* regardless of volume.

Finally, practice intending your *"Hello"* silently. This doesn't mean just sitting there and "thinking" the word. Get the same sense of *intention* as when you were using words and sound to *communicate* it. Alternate this with speaking the word, until you can maintain the strength of the *intention* when silent.

"One-way" *flows* of communication can sometimes feel depleting if *run* too long in the same direction. You may have noticed that during a publicity event or workshop that the "Q-and-A's" or more personal "book-signing" segments take place *after* an individual lectures or reads. This allows a natural *replenishing* of *attention units* that have otherwise been directed or projected *outward* for a long period of time.

You can actually balance this out in the above *processing* exercises by *imagining* that the objects are saying *"Hello"* to you, placing you at the *receipt-point*. The emphasis here should be on also spotting the object as a *source-point* of the *communication*. [At no time do we expect the object to audibly say *"Hello."*]

ACKNOWLEDGMENT PROCESSES

In two-way *communication*, there is another component to a true cycle: *acknowledgment*. The message crosses a distance from a *source-point* to a *receipt-point,* and the "receiver" *acknowledges* that the message has been received. This completes a full cycle-of-action. Another cycle may then begin.

Compulsive attempts to *communicate* result from lack of *acknowledgment.* An individual continues to *outflow* until they can sense that their intention has been received. You may have observed this in everyday life, where an individual continues to basically say the same thing until the receiver finally says, "*okay, I get it.*"

In *Traditional Piloting,* acknowledgment is essential for completing a cycle or PCL. A *Pilot* directs the PCL; the *Seeker* receives it and performs the action; the *Pilot* completes the cycle with an *acknowledgment*—usually "okay," "thank you," "all right," *&tc.* This is a critical part of *systematic processing* when it is "Co-Piloted." In *Solo-Piloting,* a *Seeker* still acknowledges to themselves (e.g. "*okay*") when an action (or PCL) is completed.

An acknowledgment is also a communication—and therefore carries its own *intention.* In the everyday life example above, a person might intend their acknowledgment to mean either, "*okay, I see what you're saying*" or "*okay, I heard you a hundred times already, just shut up.*"

Using the previous "*communication processes*" as a model of practice, let us do some exercises that concern the area of acknowledgment.

"*Imagine objects in the room saying 'Hello' to you.*"
"*Acknowledge each 'Hello' by saying 'Thank You' out loud.*"

As before, a *Seeker* can also practice this by intending a "silent" acknowledgment, once they have a sense for it with words and sound. And, as before, a sense of the actual *intention* is the emphasis of the *process.*

DUPLICATION AND REPRODUCTION

There are two main aspects of *duplication* that directly affect a *Seeker*—in *processing* or otherwise: the ability to duplicate an action (the same thing repeatedly); and the ability to copy the meaning of what is being communicated, or to even put an advanced spin on it, to duplicate someone else's *"point-of-view"* ("POV").

Ability to properly handle duplication is critical to the "spiritual well-being" of the *Alpha-Spirit*. Having experienced a very long existence, an individual that has suffered from many undesirable incidents will likely go "out-of-communication" (or "out-of-reach") with such areas, finding it unacceptable that these things should repeat. This develops into automatic-reactivity that make us reluctant to repeat anything.

There is also the factor of *imprints* on the *Backtrack* remaining from lifetimes spent in slavery—often forced to do repetitive tasks under duress. There is a lot of *turbulence* in this area for most *Humans*, and it has caused an increase of unhappiness and illness when triggered or restimulated in modern society. Of course, if *"duplication"* is *processed*, a *Seeker* can experience a relief from the discomfort associated with it.

As has been the general method in these early lessons, we'll start with the *"Willingness"* to *duplicate*. This is a light *subjective process*.

  1. *"What would you be willing to have happen again?"*
  2. *"What would someone else be willing to have happen again?"*
  3. *"What would others be willing to have happen again?"*

In this instance, it may be apparent that our systematic approach of revolving circuits is also meant to keep a *Seeker* from *running* a single *flow*-type for too long. Repetitive-style *processing* is sometimes *run* in *Co-Piloted* sessions, because ultimately we are mostly concerned with the first PCL. But for a *process* like this, a *Solo-Pilot* will assuredly not get the most significant results by continuously *running* that one *flow* alone.

"BELL, BOOK & CANDLE"

The *"2020 Professional Piloting Course"* —now collected in the *"Metahuman Destination"* volumes—occurred shortly after release of the first two advanced publications: *"The Tablets of Destiny Revelation"* and *"Crystal Clear: Handbook for Seekers."* During this time, we spent a lot of time experimenting with methods of increasing a *Seeker's "presence in-session."* The subject of *"duplication"* sprung up continuously.

In the original version used at the *Systemology Society*: a bell, a book, and a candle, are all placed on the table in front of the *Seeker.* They select two of the items and the third is put away. This validates a *Seeker's* power of choice at the start of the *process.* This may be performed with *any* two *dissimilar* objects.

A version of this now appears in our modern practice (script) of a *Formal Session,* in the *"Opening Procedures"* listed as "Control of Body and Mind In-Session." In those instructions: the items are placed within reach; or alternatively, at two points in the room (preferably on tables), in which a *command-line* for "walking between" would be inserted in the *process.* [Objects are listed as *"Item-1"* and *"Item-2."*]

"Pick up Item-1."
"Notice its weight."
"Notice its color."
"Notice its texture."
"Put it down."
"Pick up Item-2."

"Notice its weight."
"Notice its color."
"Notice its texture."
"Put it down."

This is performed repeatedly over and over again; but it is not really as simple as it seems. Each cycle must be performed just as if it is the "first time" and *not* simply as an "automated repeat" action. One of our goals with this *process* is to *"run out"* — or *"process out"* — the reactive tendency of putting repetitive action "on automatic." It also treats the tendency of *re-creating* the *past* as the *present*.

To be extremely effective and provide stable results, it is not uncommon at first to *run* this *process* in excess of 20-30 minutes, and then for as long as an hour or two. This is not how it is treated as an *Opening Procedure* in a typical *Formal Session*; but, it may be when it is first introduced to a *Seeker* by a *Professional*. It is also not uncommon to experience the entire *"Beta-Awareness Scale"* of *emotional* and *mental* states while *running* this.

As a *Self-determined Alpha-Spirit*, there is nothing inherently wrong with setting a "system" or creating a "mechanism" to do things "on automatic" *knowingly* by choice. Unfortunately, there is a lot of this built into the *Human Condition* that is experienced *unknowingly*. Therein lies the *fragmenting* factor.

It is certainly within the capabilities of a *"god-like"* being to perform the same action repeatedly without giving in to any "hypnotic effect." Properly *running* *"Bell, Book & Candle"* allows a *Seeker* the practice of breaking "hypnotic effects" of repetition — but only when seeing each performance of the cycle as a "new action," intended in its own unit of time, and not the cumulative copy of the past.

There is an advanced application of this *process* that may be applied on additional passes through this material (if it is found to be above a *Seeker's* present skill level). A *Seeker* should be well practiced with the standard ("physical") version of this as given above—and *run* it for a few cycles in the same session before applying the advanced ("mental"-"eyes closed") version.

This practice works best if you are able to *imagine* or visualize a *"space,"* rather than simply *creating* the two objects in your mind. For example, you can image the corners of a cube or a room and then add the walls, floor and ceiling. Make this large enough that it extends above, below and behind the *viewpoint* (or "POV") that you are perceiving this imagery from. [There is *no* reason to imagine yourself as a *body* in that room.]

"Create (or 'imagine') two tables in the room."

"Create 'Item-1' on one of them; and 'Item-2' on the other."

"Intend for 'Item-1' to float up above the table."

"Notice its weight; Notice its color; Notice its texture."

"Intend for it to float back down."

"Intend for 'Item-2' to float up above the table."

"Notice its weight; Notice its color; Notice its texture."

"Intend for it to float back down."

Practice the steps in alternation, as with the standard ("physical") version.

- - - -

Working through the first, second, and third lessons of the *Professional Course,* marks completion of *"Systemology Level-0."* It demonstrates all of the skills necessary for a *Seeker* (or *Pilot*) to conduct a *Formal Session* and perform *systematic processing* for additional "levels" that remains on the *Pathway-to-Ascension.*

LESSON FOUR
"HANDLING HUMANITY"

HANDLING THE HUMAN CONDITION

In this lesson we will begin using the skills we previously learned for establishing a *Formal Session*—or rather, our *"presence in-session,"* which is what makes *systematic processing* possible.

Here we will start to apply *processing* to specific areas that are necessary for a state of *Beta-Defragmentation*. This primarily means handling the *Human Condition* and *confronting* our experiences in this lifetime.

For the *processing* demonstrated in earlier lessons, we mainly encouraged simply "pushing through" any *resurfacing* or *restimulated fragmentation*. It is, however, at this level of the *Professional Course* that we begin to learn how to handle *fragmentation* directly —to start actually scraping away the layers of *imprinting* and the *considerations* about *Life* and *Reality* that have been made as a result.

This lesson is based on teachings previously given in the *"2020 Professional Piloting Course"* and contained in the text: *"Metahuman Destinations (Volume II): The Universe and Mind-Body Connection."*

[Note: if you have already attained any ultimate *realizations* as *end-points* for any particular area (from previous passes through this course material), then your practical instruction is to *"spot"* the moment it happened; alternatively if you have completed *Beta-Defragmentation* altogether, this material may be applied to *Alpha-Defragmentation* by *imagining/creating*, or apply the *processing* directly to the *"Backtrack."*]

COMMUNICATION AND PROTEST

Early on the *Spiritual Timeline* (or *"Backtrack"*), an *Alpha-Spirit* goes *"out-of-communication" knowingly* and selectively. This is part of how personal *Identity* is established. The *Alpha-Spirit* imposes or creates a certain "distance" for their *communication,* and "barriers" to their *perception,* so as not to be an integral of all *"beings"* at once.

It is in this wise that we can consider the individual's own *Spiritual Timeline* much like a *"personal identity continuum"* of an *Alpha-Spirit*—a *"ZU-Line"* that extends back to before the origins of even our "keeping track" of *time.*

As the "barriers" are originally *knowingly "Self*-imposed"—or *created* on one's own *Self-determinism*—they are usually able to be side-stepped early on the *"Backtrack"* easily with *intention.* But they are "barriers" nonetheless, and as a result, the *Alpha-Spirit* can still be "surprised" or suddenly encounter something that they weren't *aware* of, or prepared for.

Before an *Alpha-Spirit* begins to *"identify"* with *lower* material considerations for their own *Beingness* (what they truly consider *Self to Be*), they are essentially unable to be harmed—there is no "substance" for which to impact. Yet, as an early part of the formation of a basic spiritual "identity," we still find the phenomenon of *"personal preference."* This might be based on *"aesthetics"* (our perception of *beauty* and *ugliness*), but it becomes an established pattern of *acceptance* and *rejection*—and increases the likelihood for blocked or misunderstood *communication.*

When undesired *communications* and *creations* are being forcefully presented to the *Alpha-Spirit,* or when their own *communications* and *creations* are rejected by others, the individual may decide to "protest" their experience of this enforcement or rejection. This is an activity modern society is quite familiar with, so we will avoid using examples that are *too* specific.

However, when an individual attempts to *communicate* a *protest* and finds they are "blocked" on those *channels*, or toward a particular "audience" (*terminal*), they will increase the volume, or *create* something that is much more "physical" in nature and therefore more difficult to ignore. Something is *created*, but it is still likely to be *rejected*, and so the individual goes on continuing to *create* it "compulsively."

Often times, a "*protest fragmentation*" may be found at the heart of a "compulsion"—or *compulsive creation*. This is why *processing* only toward one's own *acceptance* of its existence is not enough to overcome it. The main issues must eventually be handled concerning all *flows* or *circuits* on that channel: *rejection* or *lack of acceptance* in both directions—and even the observation of others *rejecting* someone else can affect our considerations.

Communication barriers are the first factor of *fragmentation* that an *Alpha-Spirit* experiences on the *Backtrack*. Of course, because they are *Self-determined*, they are not themselves the area of *fragmentation* we target as we begin our "*defragmentation processing.*" It is "*protests*" that are the first actual source of *fragmentation* that actually lowers actual ability and *Awareness* of an *Alpha-Spirit*; in this case because of the *compulsive* activity.

The instructions for basic "*protest defragmentation*" is: "spot" (*recognize*) a specific *protest*; identify what you are *creating* or *doing* to communicate that *protest*; and identify who should have received or acknowledged that communication.

A *Seeker's* "reach" on this direct method of *processing* may be limited during their first pass through the *Professional Course* material. Many of our *compulsive creations* we carry with us as *Spiritual Beings* are protesting things and aspects of existence that are not only long-forgotten, but far and beyond what is found in *this present* manifestation of a Physical Universe ("*Beta-Existence*"). As a *Seeker* increases what they *realize*, or are *aware* of, even more becomes accessible.

PROTEST DEFRAGMENTATION

As a standard *process*, the above instructions may involve communication with a *Co-Pilot*; or a *Solo-Pilot* may wish to record their answers in a *"Flight-Log."* It is best to *run* the same area of *protest* repeatedly. Focusing on *spotting* it and describing its specifics with each *run*. This allows the *process* to naturally shift to an *earlier "protest"* in memory that is in the same area or along the same *channel*. This enables you to more fully *defragment* that entire "string" or "chain."

A. *"What are you currently protesting? Describe it."*

B. *"What have you done to communicate that?"*

C. *"Who should acknowledge that communication?"*

D. *"Imagine them acknowledging your communication."*

This same basic *processing formula* (for "Circuit-1") can also be applied to handling *"protests"* on other *flows* or *circuits*. It may be preferable to rotate these other *processes* in between *processing* your own *protests*, to keep from *running* only one *flow* excessively. The following steps apply to "Circuit-2."

A. *"What about you is someone protesting?"*

B. *"What have they done to communicate that?"*

C. *"How could that be acknowledged?"*

D. *"Imagine them receiving your acknowledgment."*

And for "Circuit-3."

A. *"What are others protesting?"*

B. *"What have they done to communicate that?"*

C. *"Who should acknowledge that communication?"*

D. *"Imagine them receiving the acknowledgment."*

As *systematic processing* becomes more intricate and specific in various areas, a *processing formula*—or basic *processing command-line* ("PCL")—is given in the text that is intended to apply to a specific *"terminal"* (a person, place, thing; something with "mass"). As a *Seeker* progresses on the *Pathway*, certain areas are found to be more turbulent than others, and are therefore targeted directly.

For example: *"What have you protested about ---?"* is directly applied to any specific area (or *terminal*) that is considered a source of *turbulence* or *fragmentation* for the *Seeker*. In *Traditional Piloting* there may be a list already prepared for a *Seeker* based on earlier sessions, or else something specific that requires additional handling. If the *answer* to this PCL is "nothing," than you simply move on to the next without pressing it further.

Without having a specialized list personalized for an individual *Seeker*, we will use some general *terminals* for our *processes*. These are: "YOUR BODY"; "YOUR FAMILY"; "JOBS"; "SOCIETY"; "HUMANS"; "LIFE ON EARTH"; "THE PHYSICAL UNIVERSE"; "SPIRITS"; and "RELIGION." Each represents its own *process*, and are used to complete the first PCL below.

The *processing formula* is:

A. *"What have you protested about ---?"*

B. *"What did you do or create to protest that?"*

C. *"Who should have acknowledged it?"*

D. *"Spot an earlier similar protest and repeat."*

A *Seeker* will generally *run* this until feeling increased relief in a certain area as an *end-point*. This type of *processing* is also helpful for those *Seekers* that at first do not find accessible answers to the more direct entry-level approach of *"What are you currently protesting?"*

When a *Seeker* is comfortable with the results of a particular *process*, or when an *end-point* is reached on what is presently accessible, deeper information may usually be "scouted" by reshaping the PCL or formula as a direction to "*imagine*" something. In this case, the wording could be changed to "*What might you protest...*" or "*What would be a communication...*" &tc. The goal being always to "shine" *Awareness* on new "layers." If, however, anything seems too out-of-reach for you right now, simply take it up on your additional passes through the *Professional Course.*

ACCEPTANCE AND REJECTION

A traditional way of scouting for (and handling) "protest fragmentation" attached to a particular "*terminal*" (or turbulent area), is PCL-alternation with a form of "*conceptual processing.*" This generally assists a *Seeker* to free up their considerations or increase their tolerance.

A. "*What about --- might you protest?*"

B. "*What about --- could you accept?*"

Enforced inhibition—when we are prevented from *having, doing* or *being* something—often produces dramatizations of "protest." This can actually manifest in two ways. For one: there is the obvious "protest" that results from being denied something, and we still want it.

There is also another side: when this "something" is denied for too long, we may suddenly *consider* that it is "bad" and that we no longer want it. But there is still a need to connect with it on some channel, so the only solution is to "protest" *against* what we wanted but were denied, just to make sure *Self* is always right.

A. "*What have you been prevented from (having, doing, being)?*"

B. "*What protest might you have about that?*"

"Protesting" is a very specific type of *communication out-flow.* Whatever that communication may be, it is quite pointedly directed at a target recipient (involving a specific "*terminal*"). Here we see one of many instances where material from our previous lesson on "*communication*" reoccurs on the *Pathway.*

The final *process* given for this area utilizes *Actualized Awareness* to "disintegrate" *fragmentation* directly. The theory behind its effectiveness is secondary to its application. However, let us consider that whatever we do not want around us, or in our space, is in some way being "protested against."

A. "*Get the concept of protesting the existence of ---.*"
B. "*Now admire the existence of ---.*"
C. "*Get the concept of you creating the existence of ---.*"

This may be run repeatedly on something you are "protesting" the existence of, even if it is not concerning a dramatization of the "marching, shouting, picketing" type we are quickest to associate with the idea. We also commonly "protest" the *existence* of certain aspects regarding "school," "work," "family," and "social governance," that we experience as *reality.* But "protest" actually concerns any strong dislike or avoidance.

To be systematically effective, *protest processing* must be handled on the direct *existence* of something—and not "something about something." If you *process* a "dirt scuff" on your shoe: the protest is "the *existence* of the scuff," and not the fact that "the shoes have a scuff on them."

This systematic approach also helps us separate our *considerations* about what a thing "*Is,*" as opposed to focusing on *fragmented associations* directly. The *fragmentation* here would make us more inclined to eventually not like our "shoes"—or even develop some other kind of automatic-response about "shoes" in general, *&tc.*

The "*admiration*" step may require some "build up" (with each PCL-cycle) to really acquire a good sense of. At first, a *Seeker* might simply "acknowledge" the *existence* of the thing; but with each pass of the step, try and apply an even greater sense of "*admiring*" it than previously.

Again, we are mostly concerned with freeing up a *Seeker's consid-erations* and *perceptive* range—undoing compulsively created *imprints* and *automated* thought. What an individual *chooses* to like or dislike as a preference is just fine so long as they are totally *Self-determined* and *Self-Honest* in their *choosing*. A *fragmented* individual only operates on a false conception of having full and total *Self-directed* control.

CHANGE AND MOTION

As a *Seeker* expects to advance upon the *Pathway,* their *willingness* to "change" must be increased. When things are difficult, there is a tendency to "clamp down" and resist change in order to avoid things getting worse. This is dramatized automatically by the *"Body"* when hurt. Essentially, the individual is trying to hold things in place—and as this starts to happen more compulsively, it becomes more difficult to change for the better.

The *Alpha-Spirit* is, itself, a *"static"* point of *Awareness* that *can* de-velop a distaste for "motion" (and to an extent, the *control* of *motion*). To some degree, there is a tendency to *unknowingly* hold everything *"still"* as a protective defense-mechanism against danger or harmful effects. Likewise, a being tends to *"keep things from going away"* as a compulsive (*unknowingly repetitive*) remedy for the experience of *loss*.

Let's shift our focus now to some *objective processing* techniques for *"Change-and-No-Change."* Each of these cycles is done five times on the same object. When you've done all three on the same ob-ject, choose a different object and do them again, five times each.

A. *"Spot the object."*

   "Place your hands on it."

   "Hold it absolutely still."

   "Decide when to let go, and then take your hands off."

B. *"Spot the object."*

   "Place your hands on it."

   "Get a sense of keeping it from going away."

   "Decide when to let go, and then take your hands off."

C. *"Spot the object."*

   "Place your hands on it."

   "Decide to move it; Decide where to move it."

   "Move it to the exact spot you had decided on, and then take your hands off."

This same *systematic processing* formula may be applied to the *"Body."* For example: using your hands to grab your right leg and *"hold it absolutely still."* Then, doing the same with the other cycles: intending to *"keep it from going away,"* and then finally *"deciding to move it"* (using your hands to lift it up and down).

Note that to *"hold it still"* and *"keep it from going away"* only appear the same at a visible level. The key difference is that, in the second one, our *intention* also includes resisting any of its efforts to move away while "holding" it there. Essentially: *our intention* is to "stop" *its effort* to "change" (or move).

This can also be practiced on other parts of the body. In some ways, it starts to establish a habit of *controlling* the *"Body"* more deliberately or *"intentionally."* Specifically in this case, an individual is *knowingly* practicing an otherwise automatic (*unknowing*) reaction to "hold" an injured or painful part of the body.

As with other *objective processing* in *Systemology*, there is also a more advanced "mental" version of this exercise. It is practiced much in the same way as the physical technique, except a *Seeker* reaches "mentally" rather than with the *"Body."* In the third step: rather than "moving" the object, the PCL is to *"make it more solid."* This is done by *intention* and *sensed conceptually.*

PROCESSING "CHANGE"

In *systematic processing*, "change" is a *channel*. It is not, itself, a *terminal*; it is a *channel* between *Self* and many potential people, places, objects, &tc. The *communications* on this *channel* involve the three basic "circuits" described in *"Lesson 3."* The following *process* demonstrates a slightly expanded version of these *circuits*.

1. *"What would you be willing to change in another person?"*

2a. *"What would you be willing to allow someone to change in you?"*

2b. *"What would you be willing to have another person change in themselves?"*

3. *"What would you be willing to have another person change in others?"*

0. *"What would you be willing to change in yourself?"*

In the above *process*, we also introduce *"Circuit-0"* for the first time in the *Professional Course*. It is not so much a communicative *flow* to others, but from *Self* to *Self*. In this case, *Self* is always the "*terminal*." In *Systemology*, this is sometimes referred to as a *"beingness postulate"* (or *"Alpha-Thought"*) when *Self directs* a decision that something *"Is"*—or about how something will either *"Be"* or *"Not-Be."*

In the end, all areas of *processing* are ultimately handling one's own *"postulates."* In essence, it is the level of *"Alpha-Thought"* that we are working up to reaching and mastering with the *Pathway*. If we were truly able to "change our mind" completely from within the *Human Condition*, we wouldn't remain in this state; so there are obviously some barriers of *fragmentation* that need to be cleared for this to be possible again.

To continue demonstrating how to *process* the area of "change," we will utilize the *"Analytical Recall"* technique introduced in *"Lesson 2."* Each circuit is treated in its own *process* containing two PCL. The PCL are alternated repeatedly until a *Seeker* cannot readily access more answers. Then go to the next *process* and do the same. After all four are completely *run*, cycle them again and see if any new answers are produced.

1a. *"Recall changing something."*

1b. *"Recall stopping something from changing."*

2a. *"Recall someone changing something."*

2b. *"Recall someone stopping something from changing."*

3a. *"Recall society changing."*

3b. *"Recall society resisting change."*

0a. *"Recall changing yourself."*

0b. *"Recall stopping yourself from changing."*

And now let's treat this more *"conceptually"* following the same basic instruction from the previous four *processes*. Notice we are using the words *"could"* and *"would"* to expand our range of free consideration.

1a. *"What could you change?"*

1b. *"What would you leave unchanged?"*

2a. *"What could change you?"*

2b. *"What would leave you unchanged?"*

3a. *"What could change others?"*

3b. *"What would leave others unchanged?"*

0a. *"What could you change about yourself?"*

0b. *"What would you leave unchanged about yourself?"*

For additional *objective processing* in this area:

"(Look around the room.) Spot something you would be willing (and able) to change; then change it."

"Spot something you find acceptable to remain the same; then leave it unchanged."

And finally:

A. *"What must be changed?"*

B. *"What must not be changed?"*

C. *"What is acceptable to leave uncontrolled?"*

D. *"What can you control comfortably?"*

As an entry-point to directly handling the *Human Condition*, we have emphasized the areas of "protest" and "change." Total *de-fragmentation* in these areas will likely take a second pass through the *Professional Course*, and possibly several cycles through the *processing* in this one lesson alone.

The ultimate *end-point* we are reaching for in this area would include the elimination of any compulsive tendencies in the areas of "protest" and "change." For example: a freedom from the *need* to change people, or prevent them from changing, in order to find them more acceptable. And, on the other side: an increased willingness to *allow* change in one's environment, or among others, without feeling a *need* to become involved.

HUMAN PROBLEMS

The type of *"problems"* handled with *Systematic Processing* are those things that tend to "hang up" in our lives—or else continuously exist without a perceivable solution. This is quite different from the kind of "logic problems" you might think of in relation to, for example, *"mathematics."*

Human Problems remain suspended in space and time as *fragmentation* because personal *attention-energy* remains compulsively fixed on them as a "problem." We are speaking of logical *"conflicts"*—or where two "things" (or *"considerations"*) remain fixedly in *opposition* to one another indefinitely.

For example: the *"problem"* of *how* to add additional rooms to your property is easily solved by logic. This would require a know-ledge and means for construction, or hiring a contractor. This is not a *real* "problem." *Real problems* would require these two things to be *oppositional*—and they obviously are not.

But could this somehow become a *real problem?* Yes. If an individual *considers* that they both *"need an additional room"* and *"have no financial means to pay for it,"* then it could develop into a *"problem"* that suspends or fixes the individual's *attention-energy* compulsively.

These two things—*"needing rooms"* and *"no money for it"* are obviously in *opposition*. The conditions cannot exist simultaneously and ever provide any kind of "solution." When not properly handled, this kind of *problem* has a tendency to suspend *attention-energy* indefinitely, which is the nature of the *fragmentation*.

Ideally, an individual will *realize* that one side or other of the *problem* has to be figured out. It is not a balanced equation to be solved as itself. Yet, the longer it remains suspended as a *problem*, the greater of an *energetic-mass* it becomes as *fragmentation*. More and more *attention* is fed into it as purely a *problem*.

In this example, the only "solution" would be to treat either the *"need for more rooms"* or the *"lack of funds."* Both of these are *considerations* only. The fact that they collide during one's lifetime into a *mass* is simply unfortunate. Either an individual would have to find an *alternative* to the *"need for more rooms"* (such as better utilizing existing space), or they would find a way to *"make more money"* or build *"more economically."*

One of the reasons *Human Problems* unfold this way is because of how much *attention-energy* they demand once they are treated as a *mass*. They have a tendency to "pull down" an individual's *Actualized Awareness* in a way that keeps them from *considering* one "side" or the other, because they only "see" the *mass*.

When the *mass* becomes a turbulent source of *fragmentation*, it may have "grown" to a state that causes the individual to feel over-whelmed (*confused*) and unwilling to *confront* it.

For example: every time the individual starts to think about "*more room*," the turbulence associated with "*money problems*" is triggered or *restimulated*, and hence they cannot think clearly about it. Then, whenever they start to try to *confront* the issue of "*money*," the worries about "*space*" become a sudden distraction.

As you can see: this individual's *problems* will cyclically continue and persist in this wise unless there is some resolution or inter-vention from an outside source. Unless the *fragmentation* is solved, however, even good fortune and charity from others will not keep the individual from falling prey to this pattern again. Their thoughts and behavior will *unknowingly* still find there way into treating this *problem* as *reality*.

Our observation of this in others also inhibits our natural desire to "*help*" others. We've seen many times that our intervention to "solve" someone's *problems* directly will not always work out. They end up in the same mess again, or it somehow seems to "backfire" on *us*. This can cause us to develop *fragmentation* in the area of "*help*" which is quite detrimental to advancement on the *Pathway*.

Therefore, in total, the main areas for *systematic processing* that we focus on in this lesson, concerning *Handling Humanity* or *Human Problems*, are: PROTEST, CHANGE, PROBLEMS and HELP.

DEFRAGMENTING PROBLEMS

The basic *processing* of *problems* begins with the PCL: "*What is the problem?*" This causes the *mass* to *resurface* directly. Then, a *Seeker* can *run* multiple cycles of: "*spot the problem*" and "*spot something about the problem*" and then "*spot the problem*" again. This allows a *Seeker* to start to control their *attention* in seeing the "*mass*" and then seeing a *"point"* of the *mass* rather than the whole. This may change how one *perceives* the *mass* thereafter.

When we say *"see,"* what we really mean is *"confront"* —and more specifically, *confront "As-It-Is."* This is something a continuing *Seeker/student* from previous lessons will be familiar with. Not only do we *process* toward *confronting* what the *Seeker* is presently perceiving as a *problem,* but in doing so, it is likely that other re-lated or similar underlying *problems* will also *resurface* to *confront* —and this is how *defragmentation* ensues.

The standard *processing* method of the above technique is:

"What is the problem?"

"What part of that problem could you confront?"

Once a *Seeker* is able to "spot" both *opposing* "sides" of the *problem,* the most effective application for these types of techniques would involve alternately *"spotting"* (or *"confronting"*) something on *each* "side" of the *problem* as the *process* is repeatedly *run.*

An *Alpha-Spirit* is an *"eternal being"* that likes to be interested in things. *Eternity* is a long time—certainly long enough for us to have our fondness of *"games"* and *"solving problems"* get turned against us. In the case of the *Human Condition,* any lingering com-pulsive interests in *creating universes* and *games,* to add richness and variety to *Existence,* is reduced to *creating problems* for ourselves. By this we are always ensured to have "something" to *do.* A *fragmented* being will always prefer a *"fragmented something-ness"* over *"Nothingness."*

In these next *processes,* we are not trying to "solve" a *problem.* A *Seeker* "spots" the *problem* and then follows the next PCL, which starts with *"imagine."* By *"imagine,"* we mean to create, invent, or visualize something. This should be an original *creation* and not an automatically (*reactively*) *recalled* event or mental image, &tc. By doing this *knowingly,* the *fragmentation* that causes the *"compulsive creation"* of *problems* comes into view, or at the very least "loosens" or "softens" for later additional *processing.*

The following three *processes* are *run* individually. If after *running* the first multiple times, the *problem* either seems more solid or the turbulence has not lessened, go to the next, and so on. They are all essentially working to accomplish the same thing and can be repeated as many times as is necessary.

"Spot the problem."
"Imagine a problem of similar magnitude."

"Spot the problem."
"Imagine something that is worse than that problem."

"Spot the problem."
"Imagine a game that would be more interesting than that problem."

A *real problem* is only brought to a *resolution* by handling it or *confronting* it *"As-It-Is."* Otherwise, whatever *fragmented* "solution" is applied to it will simply bury the original *problem* deeper and also create a new one. For example: *"borrowing money"* for the *"additional room"* only adds to the chain of complications attached to the original *"not enough money"* side of the *problem*. The individual still doesn't have enough and still has to pay for it, but now, presumably with interest. It only *avoids* the original "face" of the unsolvable *problem*.

We can apply some *processing* to this area by treating the stores of information collected (communicated) on the "circuits." A continuing *Seeker* will, by now, be familiar with how to *run* this type of PCL-series:

1. *"Spot the problem."*
   "What solutions have you had for that problem?"

2. *"What problem has someone had with you?"*
   "What solutions have they had for that problem?"

3. *"What problem has someone had with others?"*
   "What solutions have they had for that problem?"

The total handling of *"a problem"* in *processing* is not the same as an *end-point realization* on the total area of *"problems"* in general; but it is a goal. Again, this is not something we are pushing a *Seeker* to expect on the first-pass of this course. However, the *"ultimate process"* for this area is:

"Spot the problem."
"What part of that problem could you be responsible for?"

This is *run* until a *Seeker* no longer produces new answers (or until the *end-realization* for the *problem* or problems). Then *run*:

"Spot the problem."
"What part of that problem could you admit to causing?"

The ultimate *end-realization* on a single *problem* and the entire area of *problems* is: an individual is responsible for creating their own perceived problems. *Systematic Processing* is not intended as a *"therapy"* to *"solve problems"* but instead, a means of solving the need to handle or consider things *as real problems*.

ON THE SUBJECT OF HELP

"Help" is high-level *communication*. To be *"helped"*—to be *willing* to give and receive *help*—requires being *in communication*. Of course, we know that *help* is frequently taken advantage of, or even used as a control mechanism to enforce a *reality* on another. But to be *Self-actualized*, no *fragmentation* can inhibit our *willingness* to *help* and be *helped*.

Help is a difficult area for many individuals due to long *imprint-*"chains" comprised of many unfortunate experiences. Even when *help* is genuine—with "no strings attached"—it sometimes fails. When this cannot be *confronted* directly, the "weight" of accumulated failures builds up *mass* as *"help fragmentation."*

As a high-power *flow* of *communication*, "help" also has the unique ability for breaking down or surpassing what would otherwise be a "*communication barrier.*" For example: if one could find ways in which to *help* an enemy and for an enemy to *help* them, the conceptual "walls" forged with hatred and war could dissolve.

For *Solo-Processing*, it is better to emphasize the "positive" side of an area in *processing*—or at the very least, being sure to alternate *spotting* the "positive" side along with the "negatives." Doing so assists a *Solo-Pilot* to push through the *fragmented-masses* that may have developed along these channels without too easily being overwhelmed by *turbulence* or distraction.

HELP DEFRAGMENTATION

In first approaching *Help-Defragmentation*, it may be best for a *Seeker* (as *Solo-Pilot*) to direct their PCL toward general "*terminal*" areas, rather than specific examples or individuals. This expands the range of *considerations* directly within the *process* (or during the *session*). It also promotes *defragmentation* on the greater "chains" that extend farther on the "*Backtrack*" (or personal "*Spiritual Timeline*").

Systematic Processing is really meant to improve how a *Seeker* handles an area in general. The seemingly current or presently restimulated "*problems*" and "*upsets*" that are more specifically targeted in some *processes* is merely a factor of *running* those areas. If something is presently triggering activity in a certain area, that would obviously require handling before one can get total stable control over that entire area.

Much like some of the other general areas of *subjective processing* already explored in the *Professional Course* series, we will use PCL that employ the words "*could*" in order to free up our *considerations* further *in-session*. This means that there is no pressure, directive, or insistence, that a *Seeker* actually "act" on any of these *considerations*. The "*answers*" do not all necessarily even have to be logical or realistic. We are simply treating all aspects of the entire area *systematically* in our *processing*.

These "*help-processes*" are the final *defragmentation* techniques provided in this lesson. They are simple *repetitive processes* using alternating PCL. A *Seeker* repetitively *runs* the PCL, simply *spotting* (*locating, identifying, recognizing in Awareness, &tc.*) the various ways you *could help* or *would be willing to help, &tc.*

When cycling though these *processes*, a *Seeker* is often pushing through "mental barriers" of *consideration* and other *postulates* (or "*Alpha-Thought*") generated from a state of *fragmentation*. "*Answers*" may not always be immediately obvious, but they tend to *surface* in "layers." This applies to more than just *Help-processing*.

So, a *Seeker* may sometimes reach a point in the *process* where they really have to reach to come up with something and still its a struggle or they can't "find" an "*answer*" in their data-banks; but then something new "occurs" to them, and they suddenly start rapidly *out-flowing* a whole new group of "*answers*." This type of "*flash*" or "*sudden realization*" is a large part of what we are after when using these *processes*.

When listing your "*answers*" in *Solo-processing*, it is quite acceptable to simply "*read/run*" the PCL *once*, then write down as many answers that come to mind; or if you only come up with one or two and feel your mind kind of "wandering" afterward, then you might "*read/run*" again to reorient yourself. When you feel you have completed one *process*, simply go to the next.

Willingness to Help

1. "*Who or what would you be willing to help?*"

2. "*Who or what would you be willing to have help you?*"

3. "*Who or what would you be willing to have others help?*"

General Help Process

1. *"How could you help someone else?"*

2a. *"How could someone else help you?"*

2b. *"How could someone else help themselves?"*

3. *"How could someone else help others?"*

0. *"How could you help yourself?"*

Past Help (*four separate processes*)

1. *"What help have you given to someone?"*

   "What help have you not given to someone?"

2. *"What help has someone given to you?"*

   "What help has someone not given to you?"

3. *"What help have others given to others?"*

   "What help have others not given to others?"

0. *"What help have you given yourself?"*

   "What help have you not given yourself?"

To demonstrate more specific *Help-processing* in this lesson, it is necessary that we again employ a *processing formula*. Here, a *Seeker* uses the basic structure for separate *processes*, each of which may target a specific *terminal*. The suggested terminals (in chronological order) for introductory use are: "BODY," "CHILD," "PARENT/GUARDIAN," "LOVER," "TEACHER," "LAW ENFORCER," "PRIEST," "POLITICIAN," "ANIMAL," "TREE," "SPIRIT," and "GOD."

The *processing formula* is:

A. *"How could you help a ---?"*

B. *"How could a --- help you?"*

C. *"How could a --- help another?"*

D. *"How could another help a ---?"*

E. *"How could a --- help themselves?"*

What generally occurs when *processing* an area intensely—such as demonstrated with the above *Help-processes*—is that the individual's own personal *"definition"* for (or *consideration* of) a specific area—such as *Help*—changes many times. It is from the viewpoint or point-of-view (POV) of that new *"definition"* that the next layer of *"answers"* originates from. This is how a *Seeker* *"systematically"* frees their true power of choice.

Much of what is given as *Help-processing* really serves to assist in breaking down those "barriers" and *masses* that are *created* by the *Seeker*—even if *unknowingly*—which blocks their *Awareness* from truly contacting, experiencing, and therefore, *confronting*, this physical existence. It is only once these *"gates"* begin to be opened —once these *"walls"* we've *created* begin to break down—that we will *realize* we have only been sealing up our own entrapment this entire time.

- - - -

Working through this fourth lesson of the *Professional Course,* in combination with the previous three (and the *Basic Course*), marks completion of *"Systemology Level-1."* It demonstrates the first required step, an increase of *Actualized Awareness*, necessary to *actually* "improve"—which is to say, a *willingness* to "change" for the better—as the *Seeker* progresses further on the *Pathway-to-Ascension.*

LESSON FIVE
"FREE YOUR SPIRIT"

RELEASING THE SPIRIT

The previous lesson for this *Professional Course* series emphasized *"Handling Humanity"* —or else the *processing* of *Human Problems*. These areas included *"protest," "change,"* and even the subject of *"help."* It is important to handle these "surface areas" that seem to more visibly "press upon" or "interfere" in the daily experience of life, before setting our sights on higher vistas—such as we will approach in *this* lesson.

To begin this lesson, we will focus more directly on treating the *Seeker* (or *Self*) *as* an *Alpha-Spirit*. This is always the intention—because it is the actual *Self* or *I-AM-Awareness* as an *Alpha-Spirit* that we are treating with *Systematic Processing*; not a *"Body"* or even a *"Mind."*

It is the *Spirit* that we direct our *"processing command-lines"* ("PCL") to when applying our philosophy as exercises or techniques. It is *Self* that processes the command, then either performs the action or directs that command to a *"Mind"* or *"Body"* if it is called for. Both the *"Mind"* and *"Body"* are *constructs*; *Self* is Eternal.

"Releasing the Spirit" is a continuous goal as a *Seeker* progresses further on the *Pathway-to-Ascension*, and it is composed of many parts. It is not necessarily a "practice" or "technique"—or even a single area—in itself. We are continuously interested in "freeing" the *Spirit* from the trappings of a material existence.

On the one hand, we have the *considerations* of *Beingness* that an individual experiences for themselves as an *Alpha-Spirit* (or in many cases, as restricted to the *Human Condition*); on the other, we have the accumulated *fragmentation* that heavily "weighs" on the *Spirit,* and of which they are *compulsively creating* as "chains" that bind their *considerations* to lower-levels of material *Beingness* (*e.g.* the *Human Condition*).

It is a *cycle.* And in this lesson—as we start "*Systemology Level-2*"— we will explore and *process* each side of this reoccurring *cycle* directly.

SPIRITUAL BEINGNESS

In our basic state—back behind all the circuitry of "*Mind-Systems*" and "*Bodies*"—we are a single unit of *Spiritual Awareness* (or "*ZU*"). In the past, philosophers that have come close to a true understanding have associated this true "*Spiritual Self*" as "pure thought."

To say that we are "*thought*" and that everything is "*Mind*" is only a part-of-the-way-there kind of truth. It at least demonstrates an understanding that *Self* is not a "*Body.*" For some, this itself is milestones ahead of identifying exclusively with a "*Body*"; or certain *body parts* used to "*think*" with (as the expression goes).

The *Alpha-Spirit* is not composed of physical matter or energy; nor is it dependent on it for its own existence. In actuality, we are an *Awareness* with the ability to generate and observe *thought.* We also have the ability to *create energy* without requiring an "outside" *source.* A full *Knowingness* of all this is what we seek to reclaim ultimately with our *Ascension.*

There are spiritual philosophies and metaphysical traditions teaching about various "*astral*" and "*subtle*" *bodies* that also enshroud the *Alpha-Spirit.* These do exist—and we presume that they are metaphysical constructs used by the *Alpha-Spirit* in former *Universes,* or less "physically solid" (less condensed) versions of *Beta Existence.* However, none of these "subtle bodies" (as the mystics call them) are the true "pure" *Spirit* either.

In our native state, we are not located, or even locatable, in "space-time"—because our existence *precedes* the *creation* of *Universes* to be located in. The idea of *Self* being "located" in a specific spot is really a matter of *reality-agreements* for the practical purposes of, for example, participating in a "game" far more than a matter of "actual fact."

A *"Free-Spirit"* can locate its *Awareness* (as a "point-of-view") anywhere it chooses simply by *intention.* This means, when not "entrapped" by *fragmented considerations,* one should be able to be *in* the *"body"* or *out* of it, at will. One should also be able to freely *pervade* their own *"Mind"*-construct, and be in total control of it.

The traditional *"astral work"* that is found in contemporary "New Age" material differs greatly from the type of *"spirit vision"*—or *"ZU-Vision"*—that is sought with the practice of *Systemology.* Most *"astral"* work is still very much entangled in the *Mind,* even if *Awareness* is separated from a *body.* Our practices are directed at points of *Awareness* superior to all the various energy fields and subtle bodies.

"ZU-Vision" is treated more directly at higher levels of *Systemology* work—but there are many reasons why we introduce the concept earlier on the *Pathway.* Most importantly, it is a "phenomenon" that *may* be encountered early on the *Pathway*—especially during the practice of *"objective processing"* (as given in earlier lessons), particularly when one is practicing the "advanced mental versions" of those techniques.

The pure *Spiritual Awareness* or *Alpha-Spirit* is the actual YOU—your *Self*—and it is quite capable of directly *creating, operating* and *perceiving* independent and exterior to *any* "Body." However, as beings entrapped by our *considerations* for *Beingness* within *this* "Physical Universe," we are primarily dependent on an *"organic-body"* (or *"genetic-vehicle"*) in order to relay to us the sensory information of this existence.

Unfortunately, without achieving a high state of *Knowingness* during one's lifetime—from which *Ascension* may be reached—as an *actualized* state of *Awareness*, the metaphorical "gravity" (or "pull") of the *material world* and accumulated *fragmentation* (kept in all these "subtle" *bodies*), "weighs" heavily on *Self*; and it will continue to linger about looking for another similar "*body*." We do not automatically become *Free-Spirits* imminently at "death."

"ZU-VISION" AND PROCESSING

There are some *systematic processes* that prompt a *Seeker* to place their *Awareness* "exterior to" a *body*. By this, we do not simply mean "place their *attention*"—but to literally "*Be*" outside of the *body*. In actual practice, we do not mean "taking flight on some out-of-body journey"—but a stable ability to "stand outside" the *Body-Mind* systems and continue to control their operation as something altogether separate from *Self*.

This may happen earlier on the *Pathway* while *processing*, although its direct handling is reserved for upper-level "*Advanced Techniques*" (A.T.) of *Systemology*. However, we take this up now to provide a *Seeker* some sense of familiarity in case the phenomenon begins to occur; and there are some fun *processes* to *run* on this even if it has not happened.

A *Seeker* might experience the phenomenon very "suddenly" without intending to. This is a natural part of some of the *objective processing* when *run* for long periods of time. Although there is no danger in it, the "suddenness" and "unexpectedness" of the occurrence can affect later attempts to do this *knowingly*.

It usually is not the "getting out" part that causes *fragmentation* in this area, but the sudden "snapping-back-in" to a *body* that results from being startled by the phenomenon, or if the *body* is disturbed in its physical environment. This can create a *turbulent charge* on the entire subject, which of course makes later applications more difficult.

In an ideal situation, the *Alpha-Spirit* is *defragmented* and can easily move "in" or "out" without any sensation of "impact"; however, when the phenomenon occurs unintentionally, or an individual "gets out" carrying a lot of their *fragmented energetic-masses* with them, the sense of "impact" *can* be uncomfortable.

Barriers that hinder the *Awareness* of this native state and ability of the *Alpha-Spirit* are broken down little by little with *systematic processing*. This means that a *Seeker* is likely to experience the sudden phenomenon of "*ZU-Vision*" (perhaps for the first time, *knowingly*) during a *processing session*. Having some sense of this ahead of time will ease any worries and prepare a *Seeker* to handle it, if and when it does occur naturally.

In these early practices, we are not concerned with actually pulling your "normal" *point-of-view* (or "POV") of *Self*—and all of its "mental machinery" and "energy bodies" *&tc.*—out from their "usual" positions. And this is not necessary at this juncture of the *Pathway*, for the *Alpha-Spirit* (not being actually locatable itself) has an ability to establish more than one POV simultaneously, and in multiple locations.

A *viewpoint*—or "*point-of-view*"—is a "point" *from* which to "view," or else "operate" our *Awareness*. In basic terms, all you would need to do to *create* a *second* (or *alternative*) one is to decide on a *spot* (or *point*, if you prefer) and start operating as an *Awareness* from it. You may leave your previous ("normal") *Human* POV where it is for these early practices, and simply *add* an additional one.

The actual YOU—the *Self* or *Alpha-Spirit*—is always *present* (providing *presence*) wherever *attention* and *Awareness* is oriented or directed. So, *you* are never using a POV "*remote*" from *you*—but in this case, we mean a POV that is *remote* from the *body*. This is what makes common use of the term "*remote viewing*" misleading.

LOCATIONAL "POV" PROCESSING

Locational or *"POV" Processing* first appeared in the *"Imaginomi-con"* volume of the *Systemology Core*. This practice is inspired by mystical training techniques found in *Franz Bardon's* work (see the *Basic Course, Lesson-5*). Therein, *"transference of consciousness"* is described—where an *initiate* practices imagining their POV "going inside" solid masses (objects) that are separate from the *body*.

However, the original methods all seem to emphasize *flow* in *one-*direction only: the "going in" part. We can improve this practice for our *systematic processing* by using alternation: "going in" and "going out"—with repetition and fluidity.

Early experiments by research-members of the *Systemology Society* demonstrated that, initially, the best results come from practicing with "large" masses—ones that a *Seeker* is already familiar with, but which are not present in the immediate vicinity. Our standard instruction at the *Mardukite Academy* is to use "a mountain" (one which the individual is not already sitting on, if that is the case).

In *Traditional Piloting*, a *"Co-Pilot"* is not likely to have an objective reality on the *Seeker's* experience—or see whether or not the ac-tions are being carried out. To remedy this: a *Seeker* may sometimes say *"okay"* after completing an instruction—or if a bit of time has passed in silence, it is customary for a *Pilot* to inquire, *"did you?"* This inquiry should not relay a tone of skepticism, but simply to prompt additional communication.

These practices found in *"Systemology Level-2"* do *not* demand a *Seeker* to have already achieved the much higher-level goal of completely separating from the "normal" *compulsively created POV* of *Beingness* "in a body." This *process* is practiced by *"imagining"* a *secondary viewpoint*. There is no reason to imagine some *"mental body"* (or any connected "cords" or "threads") in order to *create* or *visualize "mental imagery."*

A better understanding of what we mean by a *secondary viewpoint* or *POV* is achieved by direct practice. The traditional *"Locational POV"* processing command-lines (*"PCL"*) are:

A. *"Imagine being above (a mountain) looking down on it."*

B1. *"Imagine your viewpoint moving in to it."*

B2. *"Imagine your viewpoint moving out from it."*

In this formula, a *Seeker's attention* and *presence* are oriented in the first PCL; the last two PCL (*"B1"* and *"B2"*) would be alternated repeatedly.

It is expected that most of what a *Seeker* first experiences with this *process* is purely *"imagination"*—meaning that it is *Self-created* in the absence of actual perceptions. Eventually, with practice, a *Seeker* will find that more of their *Awareness* is able to be "established" or "present" in the *secondary viewpoint*.

Recognition of "real perception" with this practice may be subtle at first, but still seem quite real. This is what we would mean by stably *establishing* a *secondary viewpoint* that is "remote" from the *body*. It is *secondary* and *remote* only to the POV that we apply to the *body* and experience with its eyes.

During phases of early progress on the *Pathway*, it is important to acknowledge any personal success or improvement in these areas by simply *ending* the *process*. To continue such practices further without a break may result in some anomaly that causes you to invalidate the gains that were made and feel like all of it was entirely imagined. Consider any "win" as an *end-point*. At the very least, return to it after a break.

Eventually, you would practice this beyond the point of *establishing a viewpoint* in order to continue practicing *"Locational POV"* at a higher-level of application. For example: using a "New Age" suggestion from an "Eastern" technique called *"journeying to other planets,"* we modified the above *process* to include practicing *"Locational POV"* with *any imagined* point in the *Universe*.

A. *"Be near (or above) ---."*

B1. *"Be inside of --- (it)."*

B2. *"Be outside of --- (it)."*

C1. *"Be at the center of --- (it)."*

C2. *"Be outside of --- (it)."*

D1. *"Be on the surface of --- (it)."*

D2. *"Be above --- (it)."*

As an extended process: after a *secondary viewpoint* is established ("A"), each set (e.g. B1 and B2, *&tc.*) are alternated a few times in sequence—working from B to C to D; and then the whole B-C-D cycle is repeated again. Although the *running* of this *process* could be essentially "unlimited," find an appropriate *end-point*.

ADVANCED APPLICATIONS

An entire volume of material—*"Imaginomicon"*—is dedicated to the theory and practice of *"creativeness processing."* The work explored thus far in this lesson falls under that category. These applications have nearly unlimited potential variations. There are many suggestions we can review from that earlier material.

For example, with our original practice of setting up a *secondary viewpoint*: once a *Seeker* has certainty on their ability using a single *alternative POV*, the practice is expanded to include multiple *viewpoints*.

To begin, before considering *additional alternative POV*, a *Seeker* would initially practice by simply alternating between a created *secondary viewpoint* (with eyes closed) and the *POV* of the *body* when eyes are open.

This would mean *"anchoring"* your *alternative POV* above the "mountain" with an intention of "keeping it from going away."

Once it is held steady in imagination, a *Seeker* should be able to re-sume that *viewpoint* (by intention) immediately upon closing their eyes. Then alternate between viewpoints, spotting something in each: eyes open, *POV-1*; eyes closed, *POV-2*; and so on, until one feels satisfied to end the process.

As an additional gradient of practice, a *Seeker* can increase their *actualized perception* enough to simultaneously *"look"* through both *POV* (with eyes open).

It will be noticed, of course, that the "volume" or "amplitude" of the *perceptions* received from the *body-POV* are considerably "louder" than the *alternative-POV.* To improve on this, even with eyes open, all a *Seeker* has to do is increase the *attention* that is dir-ected to the *alternative-POV* (to keep it from going away).

CREATION-OF-SPACE

The *Alpha-Spirit* has forgotten its own natural ability to *create* and handle *"space."* We have come to rely on the energy-matter of the *Physical Universe* and the sensory perceptions of a *genetic-vehicle* to have any sense of *"space."*

Although the concept of *imagination* and *mental imagery* is treated quite lightly in former instruction, the upper-level consideration is: to truly *create* energy, matter and various forms, there must first be *"space"* created for them to exist in. Although difficult to fathom for some readers at first, this is a concept that may actually be practiced in *processing.*

Many mystical practices treat philosophical constructs of *"space"* as "circles" and "spheres." For example: a "magician" *casts* a "circle" to mentally differentiate their own designation of "sacred space" from the surrounding universe. This is an example of dramatizing or mirroring *creation* of a *"Personal Universe."*

Using our methods, we developed a similar exercise that allows a *Seeker* to more *systematically* define the parameters of *created space* by using the philosophical construct of a "cube." A *Seeker* is likely already familiar with the idea of a "cube-like room" to designate a "*space*" as separate from elsewhere. This is far more effective in *processing* than mentally defining the three-dimensional points that compose a "sphere."

Our previous exercises may be extended further for a demonstration of this idea. Rather than use the *facsimile-copy* of a "mountain" as we *re-created* it from something found within this *Universe*, this time we want to actually *imagine* the "mountain" being within the "*space*" of one's own *created* "*Personal Universe.*" This is practiced by:

A. Creating dimensions (defining boundaries) of "*space*"; and then

B. Creating the energetic-masses or forms within that "*space.*"

Rather than closing your eyes and essentially "*recalling*" an actual "mountain" from memory in order to duplicate a *facsimile-copy* of the scene, this time we want to *create* fixed dimensions of finite *space* within our *Personal Universe* for which to *imagine* "a mountain" of our own unique design.

Of course, "*Creation-of-Space*" may be practiced cumulatively over the course of multiple *sessions*. It is dependent on a *Seeker's* ability to *imagine (create)* and perceive a single "*point*" with definitive certainty.

This exercise is then extended to include other "*points,*" a "*line*" between two points, a "*square*" composed of four points and four lines, and finally a complete "*cube.*" Eventually, a *Seeker* works up to being able to distinctly *perceive* the eight "*points*" to form a "*cube*" of "*space*" within a single *processing* "step" or PCL.

For our technique to be effective, when first starting out: it is far more important to distinctly and fully *create* and *perceive* a single *"point"* —and then even a *"line-segment"* between points, perhaps during a separate *session*—rather than rush progress and only vaguely have certainty on having *created* a stable *"cube."*

Once the "cube" has been sufficiently *imagined*, a *Seeker* may then practice *"intentions,"* such as: *"hold it still," "keep it from going away"* and *"make it more solid,"* to establish greater vividness of the *mental imagery*. [Refer to *Lesson-4* for details regarding use of these PCL in *processing*.]

If it is helpful, these *"intentions"* (*"hold it still,"* &tc.) may be applied (as PCL) to each progressive "step" (the *point*, the *line*, &tc.) along the way. Note that it has taken some Seekers a significant amount of practice to simply achieve a total *realization* of *creating* and "holding" even a single point "still" enough to perceive it "solidly."

The original *processing* steps are as follows:

"Imagine a point."

"Imagine a line stretching to a second point."

"Imagine lines stretching upwards from those two points to another two points."

"Imagine the upper points are connected to form a square."

"Imagine another set of four points connected together to form a second square."

"Imagine lines connecting the two squares to form a cube."

"Imagine that this cube is pure space."

Once a *Seeker* is skilled in this practice, *processes* (such as "the mountain") that involve *creating* or *imagining* (as opposed to *recalling*) are far more effective when handled within this *personally created space*. For example: rather than closing your eyes and suspending "a mountain" against the background scenery of your mind, you would first *imagine* a large *"cube-of-space"* and *then* the "mountain" within it.

THE "MUSTS" & "CAN'TS" OF LIFE

Part of the freedom that is recovered in *systematic processing*, is the ability to operate the *Human "game"* without obsessing over *"must haves"* and *"can't haves"* —or the flip-side of this, the *"must avoid"* and *"can't get rid of"*—considerations that seem to so strongly influence our thoughts and actions; or, at the very least, entangle much of our *attentions*.

There are times you may have noticed that when you had *really really* "wanted" something, it tended to remain out of reach; or when you have tried *really* hard to "avoid" something, it seemed to seek you out and practically land on your doorstep. This frustration is a part of the standard-issue *Human Condition*.

By *"must have"* we don't mean, for example, how a "body" *must have* "oxygen." We mean *intense "desire."* That being said: we aren't implying that you can't really like something and still obtain it, or have it. But, cravings and obsessions are born from *fragmentation*. The more *turbulent, fragmented,* or "desperate" the *attention* or *effort* given by someone, the more difficult it is to "manifest" exactly what they actually want.

Those things which we want (or don't want) are, indeed, *"things,"* which is to say *"terminals."* As such, our experience with them is based on *energy-flows* and *considerations*.

In the case of our usual (and often *fragmented*) participation with *Life*: that which we "reach" *too hard* for will create a "pressure-wave" type *flow* that will actually push the "thing" away; and when we try to "avoid" something *too hard*, we create a "suction-wave" type *flow* that will actually draw the "thing" toward us.

At an upper-level of understanding, attempts to *control* the *flow* using considerations of *"must have"* or *"can't have"* creates an energetic turbulence similar to what we encountered in *Lesson-4* concerning *"Human Problems."* Of course, in that instance (*"problems"*), two basic things in opposition are colliding with each other; here we are specifically considering attracting things we desire and avoiding the undesirable.

There are times when the *clear route* seems *"counter-intuitive"* — but, then we must remember that in most cases our *"intuition"* is likely to also be "clouded" by *fragmentation.* And the real fact is: as we progress on the *Pathway,* our dependency on what some call *"intuition"* is gradually replaced by *true Knowingness.*

True *control* of a "system" often requires only a light touch. In the case of handling *"terminals,"* we are again concerned with one's own level of acceptance. By this we mean *defragmenting* an area (or a *terminal*) to a point where a *Seeker* is *willing* to "have" or "not have" freely. This is the secret key to *"having."*

By being free of the *fragmentation* related to something, an individual is more likely to "enjoy" it without experiencing the interference of *compulsive obsessions.* In the same wise, *fragmentation* inhibits us from simply walking away from something we are disinterested in, without recoiling in revulsion and disgust.

DEFRAGMENTATION TECHNIQUES

Some of the best *systematic processes* for handling the *"must haves"* of *Life* utilize *"visualization"* or creating *"mental imagery."* A continuing *Seeker* has had some practice with this already. While some may consider this to be "imaginative-play" only—we are still *systematically processing* our *considerations.* In this case, we use *"mental imagery"* directly to do it, rather than representative words to consider an area.

When we desire or crave a thing intensely, it has a tendency to develop an *automatic consideration* in the "Mind-System." This *consideration* is that the thing has near-infinite value and scarcity or rarity. Such *considerations* conflict with other *reality-agreements* we have about the "*Physical Universe*" when it comes time to encounter or manifest the condition of *having* whatever "it" is.

The systematic solution—although this will seem strange to some readers—is to visualize "*wasting*" (and then eventually "*giving away*") that which we *want* to *have* so deeply and keep so closely. At first, a *Seeker* might visualize the item being sucked into a vacuum machine or being thrown out into space. Then you can start to *imagine* actually giving it away to others.

This not only increases our fluid "*acceptance*" about the item; it also changes our underlying *perception* or *consideration* of the thing being "so super rare that we probably can't ever have it anyway." Only after this point is reached would an additional step be added of *visualizing* "*receiving*" it from others.

On the other side of this, we have also discussed the idea of obsessively "avoiding" certain undesirable things that seem to always show up in our lives. We have a *fragmented* view of "scarcity" in our lives. The same mechanisms are in play as before, but in reverse. In this case, you would start the previous technique with visualizations about *having more* of an item. The "*giving away*" and "*receiving*" parts remain the same.

These types of *processes* are not meant to be *run* for excessively long periods; but long enough to feel the "release point" of breaking through a certain "mental barrier" that has been restricting the free-*flow* of energy on the channel (as related to a specific "*terminal*," *&tc*). The *end-point* is not so much of a *realization* as it is a certain "feeling" that some small weight has been "relieved" or has "fallen away."

This technique is not intended to actually "manifest" the condition of actually *having* something; only freeing up the *considerations* about *having* it, which is the required step (often missed) to getting what we want. To get a better sense of how this may be applied to various *terminals*, we'll take up an example.

"*Money*" is a good example to start; it is a general area of concern —and a common area of *turbulent fragmentation*. But, it is a good direct example—because, "*money*" is the equivalent of "*lifeforce energy*" that *flows* in a "human society." It is an integral "system" of "civilization"—and to operate at optimum efficiency, this *energy* must "circulate"; it must *flow*.

Using the above *process*, a *Seeker* would start by *imagining* ways to "*waste money.*" It essentially "answers" a PCL for considering what physical actions one might take to "*waste money,*" but by *visualizing* various scenes and possible events. *Imagery* should be *created*, not *recalled* from actual memory; and it can be completely ridiculous examples (*e.g.,* "tossing it into a fire"), so long as "*ways to waste money*" is answered.

Once a *Seeker* has worked with this, the additional steps of *processing* may be applied as *circuits*.

1. "*Imagine giving 'money' to someone.*"

2. "*Imagine someone giving 'money' to you.*"

3. "*Imagine someone giving 'money' to others.*"

In many ways, "forms" and "things" are the materialization of a particular "idea" or "concept." In the case of "*money,*" it is really a form of communication exchange that is used in exchange for various goods and services. This, again, makes it a perfect demonstrable example of the next *process*.

A. "*What could 'money' be a substitution for?*"

B. "*What could substitute 'money'?*"

Since one our goals is to freely accept and reject the same "thing" (as an *energy flow*), the following applies to more specific considerations:

A. "*What 'money' could you accept?*"

B. "*What 'money' could you reject?*"

Since we are dealing with a "thing" that we want (or don't want), we treat this type of *processing* differently than just the "concept" of the "thing" alone. If this were purely "*conceptual*," we would run: "*What 'about money' could you...*" which is a slightly different, but related matter, introduced in previous lessons.

When the more significant turbulence has been quieted down for a particular area, only then is it appropriate to apply other types of "*affirmation*" or "*visualization*" techniques that might directly "attract" a particular energy or *flow* into your life. These are what are more commonly dispensed in typical "*Self-help*" and "*New Age*" material, but without actually handling the underlying *turbulence* or *fragmentation* itself.

As a result, many are turned off by any new "meta-psychological" approach, because the old "*think happy thoughts*" or "*wish it, will it, get it*" or "*help me, help you, help me, sell more pop-Self-help books*" didn't produce the stable results promised. A few people may have been the richer for it; mainly the authors. *Systemology* is not intended to be viewed in this same light. Our results speak for themselves with each individual *Seeker*.

To experiment with our philosophy further, using a "*New Thought*" exercise (in the direction of "fluidity" and "attraction"), try the following visualization exercise with your eyes closed.

Imagine "clouds" of "money" around your body. Then using *intention*, "push" those "clouds" into the body. Be careful not to "pull" them from the "inside" as this only reinforces the sensation of *Beingness* as a body. After doing this several times, *Imagine* these "clouds" and use *intention* to "throw them away" or else "explode" off in the distance. Then alternate back to *imagining* and "*pushing them in.*"

If either of these "*flows*" seems more difficult—or suddenly becomes more difficult—then concentrate on doing the "other" a few more times before alternating. In the beginning of the exercise, it is easier to *imagine* handling "smaller" or "lower-grade" versions of the item (*"money"*) and then gradually improve the "value" represented (for example, "copper" into "silver" into "gold" *&tc*).

After you have worked through this material concerning *"money,"* consider applying the full cycle of *processing* given in this *"defragmentation"* section of the lesson to various things like: "FOOD," "SEX," and "JOBS" (or "WORK") for additional practice.

Eventually, a *Seeker* will want to *run* this on the actual (specific) stuff they *"really really want"* (and perhaps haven't been able to get, or blatantly *"can't have"*). The *"desires"* that run strong with an individual will be uniquely specific to them and cannot be covered fully here; but the same *procedure* given above is applied.

"AVOIDING" & "GETTING RID OF"

Having treated the area of *"wants," "must haves"* and *"can't haves"* above, the next step is to consider the other side of this: that which a *Seeker* feels they *"must avoid"* in *Life*, but can't seem to *"get rid of."*

Although we are treating all of these areas with words, at an upper-level, we are really handling the *energy-flows* of *attention* — and the *energetic-masses* that these form when *flowing* in one-direction too long. We can become so concentrated and fixed upon something that our *attention-energy* actually builds up a "pressure-like force" or "mass" that continues to press against something *oppositional* or *blocking*.

For example: when we were treating the area of *"must have,"* we began our *processing* by *knowingly* reversing the "stuck" (or *compulsively* "one-way") *flow* regarding the *terminal*, and *"wasting"* it, or *"throwing it away."* Once the *turbulent fragmentation* is handled, the *"acceptance-and-rejection"* type of *processing* simply provides greater stability (or certainty) for handling that *flow fluidly*.

In order to *systematically process* an *attention-flow* regarding the op-posite—something that won't go away, or that a person is trying to be rid of (or avoid)—the first PCL of the previous *procedure* would be opposite. This means reversing the *flow* by *imagining* ways of *"having more"* rather than *"wasting."* The remainder of the *processing*—concerning *"giving-and-receiving"* and *"acceptance-and-rejection"* is identical.

For the previous demonstration, we used *"money"* as a common general example of something *desirous* to *have.* For this opposite demonstration, we will also use something general and common that most lifeforms dislike strongly and try to avoid: *pain.*

Usually this "reflexive" or "compulsive" use of our *attention* will cause us to attract more of what we are obsessively trying to avoid and/or to experience it more vividly when it does occur.

As a quick rundown of procedure, the first PCL (of the first *process*) is: *"Imagine ways to have more pain."* The remainder of the *process* consists of: *"Imagine ways to give 'pain' away"* and *"Imagine ways to receive pain."*

It is important, again, to *create* and not to *"recall"* actual events for this type of *processing.* Once a *Seeker* has *run* this, the *processing* may be applied directly as *circuits* as given here:

1. *"Imagine giving 'pain' to someone."*
2. *"Imagine someone giving 'pain' to you."*
3. *"Imagine someone giving 'pain' to others."*

And for the considerations:

A. *"What could 'pain' be a substitution for?"*
B. *"What could substitute 'pain'?"*

And finally:

A. *"What 'pain' could you accept?"*
B. *"What 'pain' could you reject?"*

If you were to apply the final *"visualization exercise"* from the original *procedure* above, than you would *imagine* "clouds" of "pain" around you—alternating the *"push in"* on the body and the *"throw away"* steps. As an extension of the *procedure* (for both sides), a *Seeker* can then *imagine* the representation of it and alternate *running* concepts of "connection" and "separation" (or disconnection).

A. *"Get the sense of being connected to it."*

B. *"Get the sense of being separate from it."*

This is *run* until a *Seeker* no longer feels any *compulsive creation* of a one-way "stuck" *flow*; which is to say, there is no longer a *flinch*, *craving* or *reactive-response* regarding that "channel" (to a particular *terminal*). Again, this regimen runs into areas that will apply uniquely and specifically for each individual, so the best we can provide in this *Professional Course* is a series of *processes* that have the widest possible *application*.

LESSON SIX
"ESCAPING SPIRIT-TRAPS"

BEING AT CAUSE

In our previous lesson—*"Free Your Spirit"*—we began to explore the true nature of the *Alpha-Spirit*, or the *actual Self*, as a *Spiritual Awareness* capable of experiencing existence from any *viewpoint*. That we seem to have fixed our *attention* and total *consideration* of our own *Beingness* solely as the *"Human Condition"* is unfortunate. But, it does not have to be permanent.

Early on our *Spiritual Timeline*—or our sense of personal "history" as an *Alpha-Spirit* (that we often refer to as the *"Backtrack"*)—we are *knowingly* acting as an *"Eternal Being"* with seemingly infinite creative abilities. This is when we were the *most* at "Cause" over our own *Beingness*, our *creations*, and our experience of *existence*.

Of course, being at *Cause* means also *doing* or *creating* things that one may later regret. This starts a chain of *inhibition* toward future *action* and *creation*. By experiencing *guilt* or *shame*, an individual may even begin to *hold-back* their *communication* with others. Accumulated *fragmentation* like this in specific areas led to an *Alpha-Spirit* restricting its own *active abilities* and *willingness* to *reach*—increasingly *withdrawing*, as *Cause*.

In *Systemology Level-2*, our *systematic processing* emphasis is on *"doing things"*—which is to say, the *things we have done*. It is only after the area of *"doing things"* is *confronted* easily, that we systematically consider additional matters of *ethics*, *justification* and *responsibility*.

When this subject is initially raised, the first things you might think of concern only the most serious crimes and conceptions of *"evil."* But that is only one small part of our *spiritual rehabilitation*.

The truth is that once we begin to *hold-back* or *inhibit* certain areas (or *channels*) reactively or on automatic, we stop *knowingly* being at *Cause* for that entire area—and we may even miss out on experiencing the "good things" there too.

BASIC PROCESSING

We begin this lesson by digging right in: peeling back layers of what we may easily *recall* and *consider*, much as we did for the general area of *"communication"* in earlier lessons. The goal is to again *"break through"* various *barriers* of *fragmentation* that inhibits a *Seeker's* free *consideration* and *willingness*.

In later *processing levels*, we are more concerned with what an individual considers they *"should"* or *"shouldn't"* do—or else, what you ultimately decide *to do* in the future. But before this may be experienced in *Self-honesty*, the *fragmented inhibition* that underlies free *consideration* about *action* must be handled (or rather, *defragmented*).

For *systematic processing*: we are concerned with *"spotting"* (*identifying* and *noticing* something about) a particular *action*, and then observing if there is any particular *"resistance"* to thinking about it, or any emotional *"turbulence"* attached to *considerations* of that *action*.

On a higher-level of application, the *Pathway-to-Ascension* is a progressive *"rehabilitation"* of the native or original *spiritual ability* of the individual (*Seeker, &tc.*) who IS the *"Alpha-Spirit"* themselves. But being wrapped up in many layers and lifetimes worth of *restrictive fragmentation* inhibits true *Knowingness* of that state.

When this *Level* of *processing* is first introduced in *Traditional Piloting*, the *Seeker* is not required to give a verbal "answer" to a specific *"processing command-line"* (or "PCL") if they are uncomfortable in doing so. In this case, they simply acknowledge that they have *considered* an *answer*.

In *Systemology Level-2*, a *Seeker* may benefit from starting off *Flying-Solo*, and then *processing* any remaining *turbulent fragmentation* with *Traditional Piloting*. As usual, we will start off with light *processes* to open up the areas we will handle throughout this lesson.

Each PCL series (below) is cycled through multiple times as a complete *process*. To *run* these most effectively, a continuing *Seeker/student* should apply what they have learned about *systematic processing* in earlier lessons that included similarly-styled *processes*.

Willingness to Do

1. *"What would you be willing to do?"*

2. *"What would you be willing to have someone do?"*

3. *"What would you be willing to have others do?"*

Willingness to Reveal

1. *"What would you be willing to reveal?"*

2. *"What would you be willing to have someone reveal?"*

3. *"What would you be willing to have others reveal?"*

A *Seeker* may *run* the following process *Solo*, recording the data on paper (to see it "external" from *Self*), but then destroying (burning) it afterward. This allows a *Seeker* to *run* it without "worrying someone will find out" (which, itself, is another matter taken up later on).

A. *"What shouldn't others know about you?"*

B. *"Who would it be safe to communicate that to?"*

C. *"What shouldn't someone know about others?"*

D. *"Who would it be safe to communicate that to?"*

Once this area has been approached as a general area, *processing* may then be directly applied to various "terminals" ("persons," "places," "things," &tc). As a standard, we start by *processing* basic *terminals* that represent each of the "*Spheres of Existence*" (introduced in the *Basic Course, Lesson-2*). These include: "YOUR BODY"; "A FAMILY MEMBER"; "CHILDREN"; "SEX"; "WORK"; "SOCIETY"; "LIFE ON EARTH"; "PHYSICAL MATTER"; and "THOUGHT" (or "SPIRITS"). On additional (advanced) passes through this course material later on, consider also "CREATION" and "DESTRUCTION."

The *"terminals"* (given above) are general examples. Other similar (but more specific) *terminals* with *turbulent fragmentation* may be used in place of them. For example: "A HUSBAND" (for "Family Member"), "HOUSEKEEPING" (for "Work"), or "AN AUTOMOBILE" (for "Physical Matter").

This *processing formula* includes several individual *processes* that are used together to form a complete *routine* for *running* a particular *terminal* in this area of *"doing things."* A *routine* such as this is preferably handled within a single *session*. Insert the "terminal name" into the following *formula*:

A1. *"What have you done involving ---?"*

A2. *"What have you held-back from doing involving ---?"*

B1. *"What has someone else, or others, done involving ---?"*

B2. *"What has someone else, or others, held-back from doing involving ---?"*

C1. *"What would you permit someone, or others, to do involving ---?"*

C2. *"What have you kept someone else, or others, from doing involving ---?"*

D1. *"What could you permit others to find out about you involving ---?"*

D2. *"What have you held-out from communicating about ---?"*

E1. *"What could someone, or others, let you find out about themselves involving ---?"*

E2. *"What have others held-out from communicating about ---?"*

The "ultimate" *process* for this area is *run* after *all* previous *processing* given in this section. It does not apply "terminals"; it assists with stabilizing results and/or prompting *end-point realizations* for this level of *processing* as a whole.

A1. *"What have you done?"*

A2. *"What have you held-back from doing?"*

B1. *"What has someone else (or others) done?"*

B2. *"What has someone else (or others) held-back from doing?"*

C1. *"What would you permit someone (or others) to do?"*

C2. *"What have you kept someone (or others) from doing?"*

D1. *"What could you permit someone (or others) to find out about you?"*

D2. *"What have you held-out from communicating?"*

E1. *"What could someone (or others) let you find out about them?"*

E2. *"What have others held-out from communicating?"*

PROCESSING "INVALIDATION"

Ultimately, at the basic core of the matter, *abilities* of an *Alpha-Spirit* can only be weakened or "lost" by their own decision. These kind of decisions are sometimes known as *"postulates"* or *Alpha-Thought*. They are the *Alpha-Spirit's* prime decision for something *"to be"* or *"not-be"* what it *Is*. This includes an individual's own *Beingness*, their own "sense" of *Self*—and decisions about what *"Self"* ultimately *Is*.

Although the *actual decision* rests with *Self*, there are many ways in which an *Alpha-Spirit* might be influenced to make such decisions. In order for *Self* to *Be* "less" than what *Self* actually is (in its true native state), it must, by definition, be *"invalidated."*

The originating trigger or restimulation of "*invalidation fragmentation*" is usually from someone else, or others. It is only when these criticisms or opinions are, for whatever reason, actually *agreed* to as *reality* (made "one's own") from a higher-level of thought (or *postulate*) that they will affect *Self* (the *Alpha-Spirit*).

This *fragmentation* prompts an individual to start *validating* the *invalidation* and therefore make it "true" for their experience of existence (or *reality*). In common terms: it becomes a "*self-fulfilling prophecy*" of sorts. The individual now *fully* "*believes*" themselves *to be* "incapable" of an *ability*, and so essentially becomes so.

The proper way to develop, enhance, or regain, an ability, requires *validating* one's own successes—even when they are only small gains. Eventually, we want to be able to withstand *invalidation* from other outside sources as well; to *confront* their existence without *agreeing* to them ourselves. An *actualized* individual could face all the criticism in the world and be unaffected so long as they do not consider *Self* as "*less.*"

A professional may make many mistakes, but the skill level increases when these are not used to *invalidate* the many accomplishments that have taken place along the way. A professional in sports continues their improvement by *validating* the many goals, hits, or scores, they have made in their career. They do not *invalidate* themselves just because of a critical miss—in spite of the "boos" and ridicule from others.

In a state of *fragmentation*—outside *Self-Honesty*—the *Human Condition* maintains a quite "fragile" sense of *Self-worth* and *certainty*. The less *actualized* an individual is, the more easily their thought processes and actions may become "unbalanced"; the more easily they may be led to *invalidate* whatever slight confidence they have. Our goal is not to run from or avoid *invalidation*, but to emphasize *validation* of the "good."

On our long journey down the *Backtrack* of existence, we have experienced a great many things on *both* sides of this area—and hence our *processing* must treat *considerations* derived from *both* sides: because we have all participated in *invalidating* each other at some point or another. This even led us to being more susceptible to its *effects*, ourselves.

One reason we do not want to act in "avoidance" of *invalidation*, is because operating in such a manner often includes "*holding-back*" as a mechanism. To avoid any possibility of *invalidation*, or situations where one may encounter it, the individual holds themselves back from reaching into entire areas of existence.

The following two basic *processes* are *run* to explore *considerations* on the subject of *invalidation*.

A1. "*How could you avoid invalidation?*"

A2. "*How could you attract invalidation?*"

B1. "*How could someone else (or others) avoid invalidation?*"

B2. "*How could someone else (or others) attract invalidation?*"

This next *process* applies "*analytical recall*" to the "*circuits.*"

1. "*Recall invalidating someone.*"

2. "*Recall being invalidated.*"

3A. "*Recall someone else invalidating another (or others).*"

3B. "*Recall someone else invalidating themselves.*"

0. "*Recall invalidating yourself.*"

Now we apply a combination of traditional "*New Thought*" and *Systemology* to start handling specifics. The first PCL is:

A. "*What might you be invalidated for?*"

Once the *Seeker* has an "answer" that may be *processed*, the following two PCL are *run* until some sense of "relief" or "release" is experienced. Then use the first PCL (above) to locate and *run* another *invalidation*. "Invalidation" means "making less of." In this *process*, we are focusing on *imagining* the "opposite" of *invalidation* taking place: *validating* an individual's true *Spiritual Beingness* as they are (or appear to be).

B. *"Imagine someone else validating you for having it."*

C. *"Imagine someone else having a similar disability or weakness; then imagine yourself validating them for having it."*

Running processes for the conceptual areas of *"criticism"* and *"judgment"* is closely related to *"invalidation."* Therefore, consider tolerable *"acceptance"* and *"rejection"* for *all three areas*. These may be run consecutively (one after another, each concept applied as its own *process*). Use the concepts of "INVALIDATION"; "CRITICISM"; and "JUDGMENT" to complete the PCL below. [This "acceptance/rejection" style of *systematic processing* is introduced in *Lesson-4* of this *Professional Course* series.]

A. *"What --- could you accept?"*

B. *"What --- could you reject?"*

"HOSTILE-ACTS" & "HOLD-OUTS"

An entire volume of the *Systemology Core*, titled *"Way of the Wizard,"* is dedicated to the subject of *"ethics."* In this present *Professional Course* series, we simplify the subject and emphasize only what directly pertains to *processing*. It is necessary for completing *"Systemology Level-2."* It is not, however, an area that many *Seekers* particularly enjoy—and thus we begin to see some of them fall by the wayside of the *Pathway* around this point of progression.

The development and use of our *Systemology* is possible so late in the *"game"* of our *Spiritual Existence*, because we have only now been able and willing to fully *observe* and *understand* the patterns of our *thoughts*, *behaviors*, and ultimately the *roles* that we play out, in each consecutive experience of a lifetime. Some of what we have experienced is difficult to "face" —difficult to *confront "As-It-Is"* —and yet, without doing so, we restrict ourselves from having total access to our own *Cosmic History.*

In actual truth, as *Spiritual Beings*, we have been involved in *games* of "conflict" and "domination" for a very long time. We have experienced, again, *both sides* of most everything by this point in our existence—even if we have *blocked-out* or *blacked-out* much of this memory. But memory exists in "sequences" and "chains" —and our unwillingness to *confront* something, *blocks-out* entire areas of *Knowingness* and *ability.*

We take a systematic approach to considering the mechanisms that are attached to committing a *Harmful-Act*, which is to say, an action that directly harms *someone* else—or some other *"Sphere of Existence."* This is an important area to *process* prior to any higher-level *considerations* of *"ethics"* or *"moral justification."*

A *Hostile-Act* or *Harmful-Act* is the start of a systematic sequence that generally results in *fragmentation*. First, there is the "act" or "action" itself. Then, the mind naturally considers *"motivation"* for that action and others similar to it. For example: when one is struck by another, there is a tendency to want to *"balance"* that action, and one *considers* that they have a *"motivation"* to strike back.

To extend this example: an individual is likely to claim that the "harm" *they have done* is *"motivated"* by the "harm" *done to them.* The *Alpha-Spirit* begins to *"postulate"* this as *reality*; the Mind-System follows by correcting its *perceptions* accordingly. But, by this insistence on manifesting a "balance" of "harm" we find the individual getting entangled in a *Spirit-Trap* that it didn't see coming.

Some traditions recognize a *karma*-like mechanism at play in the Universe. In most cases, however, this is usually considered as something mediated by an "outside" or "other-determined" source. The actual fact is that we impose this *karma* upon our own *reality.* And it systematically plays out like this:

Our insistence that harm "must be" balanced then leaves us with our own "unmotivated" *Harmful-Acts* that "must be" balanced by *future* "motivators." This means receiving harm in the *future* that one feels they, themselves, deserve (to balance what they have done). What's more: having committed a *Harmful-Act*, the individual starts to *"hold-out"* (*withdraw*) their communication with various individuals, groups, or even entire areas of existence.

This whole "systematic" sequence may be taking place quite *unknowingly*, because the original *"postulate"* for such automated-mechanisms to exist were put into "play" by *Self* as a *reality-agreement* long ago. But before we begin *processing* these areas, let us consider how an individual gets entangled in such *Spirit-Traps* in the first place.

"SPIRIT-TRAPS" AND "REALITY"

The subjects of *"Reality"* and *"Existence"* are approached somewhat indirectly all along the *Pathway-to-Ascension*. Our understanding of what *Is*, and the nature of our own participation with *Reality*, improves cumulatively the whole way.

Participation in this *Shared Universe* is quite similar to our experiences of the many that preceded it. This *Reality* is a "shared illusion" *created* by the *Alpha-Spirit*—the individual themselves—that is experiencing it. This is always the case, even before an *Alpha-Spirit* "shared" their *created illusions* with others, and simply *created* in isolation for one's own *Self* as a personal "Home Universe."

Although the *Alpha-Spirit* is the ultimate *creator* of their own *Reality*, there is a separation or fragmentation inherent from the beginning in order to experience that *created reality* or "illusion." Without this separation, an individual would have "total identification" with their *creation*; and in a *Shared Universe*, it would result in a "total identification" with everyone else. There would be no sense of "individuality"; there would also be no continuation of a "growth-pattern" left unattended.

Communication is an underlying factor in establishing and maintaining *reality-agreements* concerning a *Shared Universe*. To maintain "individuality" we must see the whole as a system fragmented into parts, which may then communicate with each other across a perceived distance. The continuous experience of any *Reality* (or *Universe*) is maintained by the continuous communication among those *sharing* it.

In the most material sense, *communication* is a motion, action, or relay, of a particle (or data) from a *source-point* to a *receipt-point*, across some distance of space. It is such motion that gives us a sense of "*sequential time*," but that is not our present concern. What we are concerned with is the fact that a communication means, in essence, *duplicating* or *copying* something at a *receipt-point* the same as it exists at the *source-point*.

What we consider "*Existence*" or a *Universe* is really the continuous communication of *reality-agreements* among all those concerned. Each individual is essentially *creating* and *duplicating* the communication from their own *viewpoint*, within their own *Personal Universe*. The level or degree of actual *duplication* is reflected in the level or degree of "synchronous exchange" (or "sameness") experienced by all those sharing it.

In previous lessons, we have treated "communication" as spoken messages; but it is also reflected in the "actions" one takes with others and their environment. Actions *are* communications. They involve a relay of intention; and they are governed by the same element of *willingness* that affects our *reach* and *withdraw* in other areas of life, and in other forms of communication.

Duplication is a critical component of a *Shared Universe*. In terms of communicated action: let us consider that whether you "hug" or "harm" an individual, at some level, there is a "*mental image*" of that *reality* (and a *duplication* of it) communicated and shared between all parties (or "*terminals*") involved.

What we essentially mean here is that: in a *Shared Universe* (or *Reality*), an individual *Alpha-Spirit* is still *creating* from within their own *Personal Universe*. However, those *creations* now include a *duplication* of shared experiences. Both the *source*-role and *effect*-role are *created* as a *reality* within the *Personal Universe* of each individual *sharing* the interaction (or communication). In light of this, a *Seeker* may better understand why there is an emphasis on "circuits" in *systematic processing*. We are *creating*—albeit *duplicating*—and *recording data* for *all* interacting *viewpoints* (*Personal Universes* of other *Spiritual Beings*) of a shared experience as part of our own *reality* (our own *Personal Universe*).

An *Alpha-Spirit* is already experiencing some degree of *fragmentation knowingly* by engaging in a *Shared Universe*. This is what allows for a simultaneous "*individual*" and "*shared*" experience. The *duplication factor* allows one to get a sense of, or feel, *both* sides of a communication—and even to *duplicate* the "opposite" role. This only works out for our benefit when interactions are those that are desirable to *both* parties.

We are in constant interaction—or communication—with others and our environment every day of our earthly lives. *Imprinting* from all of these interactions is not necessarily *persistently* and *compulsively created* as part of our more permanent *reality-agreements*, and therefore may not all be a significant source of *spiritual fragmentation*. The primary factor on this is what we are *knowingly willing* to confront and *accept*.

"Do unto others..." has existed as a "*Golden Rule*" for thousands of years, but failed to perfect the *Human Condition*. We are interested in the entire experience from someone else's *viewpoint*, in addition to the actions themselves. It is not a matter of whether you would like it or *accept* it from your own *viewpoint*, but whether you would like (or *accept*) it from the *viewpoint* of the *effect*-role (the "other person"), meaning *them* (as the individual *they* are).

The real *fragmentation* begins to accumulate (and *Awareness* is increasingly entangled) when an individual becomes *unwilling* to experience the *effect* they have *created*—meaning, of course, *unwilling* to *confront* the *viewpoint* of the opposing-role. Keep in mind that we are, again, referring to something that an individual is *unwilling* to *confront* and experience, that they themselves are *creating* (or *duplicating*) as *reality* for their own *Personal Universe*. And therein lies the rudiments of a *trap* for the *Awareness* of an *Alpha-Spirit*.

SOME BASIC TECHNIQUES

To start off lightly in the areas we have been discussing, alternate the following PCL:

A. *"Recall a time that was pleasant for both you and someone else."*

B. *"Imagine the experience from their viewpoint."*

Using the above *process* as an example, place an emphasis (for "Step-B") on "getting a sense of" or "feeling" things from other people's *viewpoints*. This may also be practiced as an "*objective process*" when engaging in pleasant social interactions. The goal is to "*imagine*" or "*duplicate*" the senses or impressions perceived from another *viewpoint*. For example: how *you* might look or sound to *someone else* as you talk to them.

When we speak of *Harmful-Acts*, we mean when someone else was harmed. It does not matter what the circumstances for it are —whether intentional or accidental, whether in *Self-defense* or even to protect others; if you are *unwilling* to *confront* the action, an experience of its *effects* may be waiting in your future.

In *systematic processing*, such *effects* may be "*run out*" (or "*processed out*") so that an individual is no longer *unknowingly* and *continuously* maintaining that *fragmentation* on their ongoing "*life-track*."

The most basic solution is simply to *knowingly confront* things in *processing* by using *imagined* representations.

Much of this *"karmic" fragmentation* lies dormant for long periods of time, which is why we seldom will experience any instantaneous "repercussion" of our actions. The "balancing" *effects* generally manifest at times and in ways that are quite far and removed from the original *imprinting incident*. As a result, this "mechanism" is not even an effective "learning tool" to properly steer an individual to a higher ethics.

As we begin to *process* more specific or direct examples, it is best to start with lighter experiences. In this case, we want to locate (or "spot") a specific instance in memory where you caused someone else harm. It does not have to be a particularly significant *Harmful-Act* at first, but should be something which you later regretted —such as "hurting someone else's feelings," &tc.

The most basic method of *confronting* the action is to *imagine* the experience from the other person's *viewpoint*; and getting a sense for how they felt at the time. In order to achieve any kind of "release" or "relief" from this exercise, you may need to alternate between the *viewpoints*: first spotting your actions from your own *viewpoint*, then spotting the event and sensations experienced from the other *viewpoint.*

If this *process* makes things seem more "heavy" or "solid"—as opposed to a sense of "relief"—then it is likely that any *turbulence* or *fragmentation* is tied strongly to a similar type of incident that happened prior to it. If this earlier incident can be located, then the same *process* (above) is applied to *that* action or event.

In the case of a *Harmful-Act* where the "victim" was not present (such as "vandalism," &tc.), a *Seeker* would *"imagine"* the scenario (and "feelings") that the other person *might* have experienced upon its discovery. Progressively work your way through whatever is easily recalled for this first pass through the materials.

While we have emphasized "out-flow" (Circuit-1) of the *Harmful-Act* itself, remember that *all viewpoints* are *duplicated* from an *imprinting incident*. This means that a "victim" may also "pull in" (Circuit-2) *fragmentation* of the opposing-role (and its "karma") as their own *reality*, by also not *confronting* it *"As-It-Is."* This is what causes individuals to *"dramatize"* (or "act out to others") what has been *done to* them.

The same techniques (previously described) are applied, but this time *running* the *processes* by *imagining* the "attackers" *viewpoint* as they commit the *Harmful-Act*. This sometimes stirs up more *emotional turbulence* than when treating one's own actions (Circuit-1). In this case, a "relief" or "release" point occurs when a *Seeker* can easily *confront* the action without any particular *compulsive* desire to do it themselves.

ESCAPING THE TRAPS

It is important to note that these *"karmic"* mechanisms that we speak of are *not* the *only* reason that things happen. Just because someone acts against you does not mean you automatically deserved it as some kind of long-running tab of retribution. And while all "actions" have a certain *cause* and *sequence* behind them, everything that happens is *not* you pulling some cosmic destiny in on your reality. Things can *just* happen.

However, if there *is* a particularly *turbulent* area that persists in spite of your basic efforts (as given in the previous sections) to *confront* it *"As-It-Is,"* then this "trap-type" of *fragmentation* may indeed be (at least partly) *"in play."*

When *processing* these areas—particularly when *Flying-Solo*—it is preferred to focus on the *out-flows* (Circuit-1) regarding what *you have done*, because the greatest gains are achieved when *processing* toward *"being* at *cause."* Of course, a *Seeker* must also *confront in-*

flows or what has been *done to them* (Circuit-2); but *running* this too long (without an alternation with *out-flows*) will tend to overemphasize *"being* at *effect."*

The goal of *Systemology Level-2* (in combination with *processing* from previous *levels*) is for a *Seeker* to release themselves from the heavier "energetic burdens" that they carry as a *Spiritual Being* — that which is most accessible to *process* at this stage of the *Pathway*. We mean, of course, the "unraveling" or "dissolving" of *energetic-masses* that entangle an individual's *Spiritual Awareness* into the fixed solidity of *fragmentation*.

"Withholding" things is one way our *Awareness* becomes entangled and unavailable to us. By this, we don't necessarily mean simply not saying or doing something; but when *attention* must be actively applied in order to *restrain* one's *Self* from such, then we are *"holding-back"* our *Self (unknowingly)* in other ways too.

A *"hold-back"* on action and ability generally begins with a *"hold-out"* on communication. An example of this might be to *"hold-out"* on one's true opinion of something in order to spare someone else's feelings. Another might be to *"hold-out"* sharing something that would be socially inappropriate or unacceptable. Finally, and most critically, there is what we *"hold-out"* communicating out of guilt or fear of punishment.

In a *Shared Universe*, all of these are examples where an individual goes *"out-of-communication"* with other *"terminals"* and *"Spheres of Existence."* As demonstrated throughout the previous lessons of this *Professional Course* series, this is the first factor that leads to increased *fragmentation* and various difficulties maintaining true *Self-determinism* and a *Self-Honest* experience of existence.

As an exercise in *processing out-flow*, the standard practice in *Mardukite Zuism* and the *Systemology Society* is for a *Seeker* to write "confessional letters" (while alone) — to see the events as separate from *Self* — and then immediately burn them. The practice of "confessing to another" (in *Traditional Piloting*) *does* have spiritual value, but it can also be easily abused or mishandled (and is not covered in this present lesson).

If the "confessional letter" does not provide a sense of "relief" or "release" with a particular incident, terminal or area, there is another factor that may be involved.

The *turbulence* attached to various *Harmful-Acts* increases with the amount of *attention* that an individual places on it "internally" to "keep it in check" (so to speak). This is most critical with those things we worry about *"someone finding out."* A lot of *Awareness* is suspended or fixated on *those* things—and that intense "internalization" of our *attention-energy* also has a tendency to "pull-in" a lot of what we "don't want" in our lives.

One of the reasons this area is so critical is that: similar to how the *fragmentation* of an *"imprint"* might be restimulated by a certain *"facet"* in our environment, the *energetic-turbulence* associated with a *Harmful-Act* may be stirred up reactively (automatically) when someone *almost* discovers one of our *"hold-outs."*

This "fear of discovery" causes *attention* to "invigorate" or "validate" that area again with more of our *Awareness*. And this is generally instigated or caused by an "outside" *other-determined* source. The solution is to *confront* the thing *"As-It-Is"* with high-powered *Awareness* in *processing* on one's own determination, rather than *withdrawing* from it and feeding it with low-level *attention* each time it gets restimulated.

We address this now at our present gradient of the *Pathway* because: should anything similar to what we are describing get restimulated in any *processing*, it must be handled (with *processing*) before any stable gains or further progress is possible. The simple fact that some "thing" *suddenly appears* in *processing* generally means it is a source of *fragmentation* that requires getting under one's own *Self*-control.

REACHING FURTHER

This lesson marks the completion of *Systemology Level-2*. This is a critical checkpoint on the *Pathway* for *Seekers*. This is also a point where a *Seeker* might cycle back to the beginning of these *Professional Course* lessons and *process* their way through a second pass of all the materials presented up until now.

Although the written materials (for this level) may end here, the point to which a *Seeker* can reach with the existing *processing* given may not have been attained with their first pass.

Ideally, a *Seeker* that has fully completed *Systemology Level-2* will have a much greater and more stable certainty on "*Being* an *Alpha-Spirit* that is *having* a *Human* experience." This certainty should be at such a level as to prevent a *Seeker* from being so easily "trapped" in the "problems" of the *Human Condition* ever again.

By this, we do not mean that a *Seeker* will have completely "broken free" of the *Human Condition* at this juncture of the *Pathway*. They are still likely to experience some emotional fluctuation with daily life. But, they are not likely to become so deeply entangled in it; they have a better understanding of how the *Human* "*game*" is played in this world—and how to handle it well enough to continue their progress on the *Pathway*.

The final exercise given for this *processing level* should assist with these goals. It covertly emphasizes that you—as an *Alpha-Spirit*—are not *actually* "located" *anywhere*. It is simply part of our "*game*" in this existence—to *pretend* to "be" in the "locations" from which we *perceive* or *operate*.

The PCL for this exercise is:

"Close your eyes; spot places where you are not. Spot many places."

We say that our goal is "covertly" embedded in the PCL, because by "checking" that you are "not" somewhere, you tend to put *attention* on it to *look* at it—thereby transferring (or "projecting" if you prefer) some degree of personal *Awareness* to that location (which is "*exterior*" from the "*body*").

The PCL does not limit *where* these "places" are. They are "spotted" by *attention* with eyes closed; they do not need to be considered relatively "nearby." This practice is repeated until some sense of *exterior* "ZU-Vision" is achieved. At this *processing level*, these perceptions do not have to be very vivid or accurate, so long as a *Seeker* feels they have achieved an "improved" or "increased" sense of this spiritual ability.

LESSON SEVEN
"ELIMINATING BARRIERS"

GAMES AND BARRIERS

Participation in, and experience of, the Physical Universe ("*Beta Existence*") is likened to a "*game*" in our *Systemology*. For philosophical and practical purposes, we apply *game theory* concepts to our *systematic processing* with regards to a *Seeker's* experience of "*Universes.*" As such, an intellectual pursuit into *game theory* is of increasing interest to a *Seeker* as they progress along the *Pathway* (particularly at upper levels).

Perhaps the first thing that you should know about "*games*" is that they consist of certain "*rules*" or "*reality-agreements*" in order to functionally exist. In addition to an assigned "*space*" or "*game-field*" (or else, a *Shared Universe*) there are very specific *considerations* that define the *reality* of "*abilities*" (or "*freedoms*") and "*purposes*" (or "*goals*").

Reality-Agreements also define the "*barriers*" (*boundaries* and *obstacles*) that allow us to actually *have* a game. Unlimited freedom; no game. Unlimited barriers; no game. A lack of any purposes or goals; no game. So, for us to *have* any sense of a "*game*" in our lives, there are ultimately "*barriers*" in place that affect (or restrict, on an apparent level) our perceived *freedom* of play—meaning our *abilities* and *Knowingness.*

An *Alpha-Spirit* participates in *games* because it is something to *do*. We have all existed for a very long span of perceived *time* and have experienced countless different *Universes* along the way. We have also all *played* countless different "*roles*" during the course of our *Spiritual Existence*; made decisions based on perceived "*goals*" in each of these lifetimes. All of these aspects are considered within the domain of our *game theory.*

While our application of *game theory* reoccurs frequently in later *processing levels*, our present concern (when first entering *Systemology Level-3*) is specifically the subject of *barriers*. This lesson directly continues from the material given in the previous one, and its invitation for a *Seeker* to *"reach further."* We will begin with some *"upper-level" Systemology philosophy.*

FRAGMENTATION AND BARRIERS

When first engaging in *Shared Universes* (or *Games Universes*), an *Alpha-Spirit* *"agrees"* to the *"reality"* of certain *barriers* in order to participate in and experience a specific *"Game of Life."* These *"reality-agreements"* make the *barriers* seem more *solid* than they *actually* are, and seem more *real* than the very *agreements* and *considerations* themselves that compose them.

During one's lifetime, an individual gets to believing that there are even more *barriers* than there are—and this is even the *esoteric* basis for dividing our perception of *ALL-Existence* into a series of *"veils"* or *"gates."* For example: an individual believes they cannot *perceive, know,* or even *think* about certain things, because it somehow *"goes beyond"* or *"exceeds"* a certain *barrier*—even though that *barrier* does not actually exist.

Fragmented and low-*Awareness* states (that are typical of the *Human Condition*) allow an individual to falsely believe that the *barriers* and other *"game-mechanics"* are superior (or more solid/real) than the *agreements* and *considerations*; but *"games"* and *"barriers"* are really a *"product of" thoughts*. The *agreements* that compose the observable solidity of this *Universe* are really just a matter of *"convenience"* in order for us to have any kind of *"shared" reality* with others that are also experiencing this *Universe*.

The *"mechanics"* are simply solidified *"reality-agreements"* that provide an internal logic or consistent pattern for the *game* or *Universe*. Otherwise, consider three friends walking down a road: spontaneously, one turns into a tree; one disappears into a black-hole; and one simply sinks through the pavement, right through the earth, and into outer-space. Such *"randomness"* does not manifest on this planet because of agreed-upon *game-mechanics* inherent in the makeup and design of *this* specific *Universe* (*"Beta-Existence"*).

We *agreed* to the *game-mechanics* (of the *barriers*) so long ago that we have forgotten about it. *Fragmentation* leads us to believe that we are completely under the *effect* of these *"mechanics"*—and that this *Physical Universe* (or *Beta-Existence*) is somehow superior to our own true existence as *Alpha-Spirits*, when it is not.

But, these original *reality-agreements* were *"postulated"* into *existence* from a higher-level of *"Alpha-Thought."* They become quite solid by comparison to the level of *thought* and *consideration* available to the standard-issue *Human Condition*, which generally operates from a *viewpoint* "below" the level of those *agreements*. Thus, the "power" of *Human Thought* does not produce the same solid level of *effect* on one's environment.

It becomes quite apparent (from within our philosophy) that *fragmentation*, itself, is the only *real barrier*—and it is all that allows an *Alpha-Spirit* to get overwhelmed (or overpowered) by *Beta-Existence*, and believe *Self* to *be* something less than it actually is. This is what restricts an *Alpha-Spirit's* freedom and ability to fully "act" *within* the framework of the *game-mechanics*, when *Self* created and agreed to them in the first place.

In a state of *fragmentation*—such as the *Human Condition*—relatively "newer" *considerations* have less impact on the *mechanics* of the apparent *"Objective Universe"* than those *reality-agreements* made as *"Alpha-Thought."* The "original" *agreements* are more "solid" in their apparent *reality* than "newer" *beliefs* and *considerations*. It is in this wise that an *Alpha-Spirit* became the *effect* of their own *creations*.

When an individual *"thinks"* or *"considers"* from a *viewpoint* within the *Human Condition*, they are running up against the very *game-mechanics* of *Beta-Existence*—and by this we mean very specifically: the high-level *agreements* an *Alpha-Spirit* has formerly made about *space-time* and *energy-matter* in this *Physical Universe*.

Some of the most basic *systematic processing* techniques (such as the *"objective processing"* exercises found throughout this *course* and in a *Formal Session*) are strongly effective because they put a *Seeker* into such *clear communication* with the *Physical Universe* that they can more easily reclaim the *certainty, spiritual power,* and *creative ability* of their own original *"Alpha-Thought"* (*agreements* and *postulates*).

What is described within this section of our lesson is essentially the actual theory behind "opening procedures" and various techniques for establishing "presence in-session" used in formal *processing.*

For example: when a *Seeker* repeatedly (and *knowingly*) *"looks at"* and *"contacts the solidity"* of a *wall* that is in front of them with total *Awareness*, they really *"see"* the *wall* that *is* there *"As-It-Is"* (and often for the first time). [Refer to *Lesson-1 "Increasing Awareness,"* section titled: *"The Wall."*]

In essence, there is an "upper-level" *realization* available that *Self* has *created*, and is *agreeing* to, the *mechanics* of the *barrier* on a continuous and compulsive basis. Once this is *recognized*, a *Seeker* can then begin to practice regaining the true "power" behind their *considerations*, and again have any high-level *control* (*"Alpha-Thought"*) over the actual *reality-agreements* and *mechanics* of a *Universe*.

PROCESSING "BARRIERS"

The *mechanics* of a *Shared/Game Universe* become such a point of inherent personal *fragmentation* that they act as apparent (or even visible) *"barriers"* —at least until the individual is again *able* to *be* free of them. It is not the *agreement* to have *barriers* and *games* that is problematic; the *fragmentation* occurs when the *Alpha-Spirit* is no longer *aware* of these *agreements*, yet continues to compulsively maintain their *reality*.

There are many times when it seems like certain applications of *systematic processing* shouldn't be necessary, since it should not be altogether difficult to "change your mind" about things. And if we could easily get an individual (entrapped within the *Human Condition*) to *actually* "change their mind" at an upper-level of *Alpha-Thought*, then indeed, *processing* wouldn't be necessary. But that is not the general experience of *Life*.

Systematic Processing is intended to assist a *Seeker* in eliminating the *barriers* of their own *"blindness"* or *"spiritual occlusion"* —meaning their *unreality* or *unknowingness* on the *reality-agreements* they've previously *agreed* to.

In the end: a *fragmented individual* is working against themselves in this *Universe*—working against their own former *agreements*. This only furthers the *solidity* of, and entrapment of a personal *viewpoint* within, this *Universe*. Using *force* against *force*, *energy* against *energy*, only strengthens the *reality* of this *existence*; whereas true *Alpha-Thought* requires no *"force"* or *"energy"* to *"postulate/create"* something into being.

Our previous example regarding our *"wall"* exercise is quite apt, because it also represents a *"physical barrier."* After making *agreements* for the *reality* of *solid-matter* and a dependency on *viewpoints* attached to *physical eyes* (of a *body*), the *wall* represents a *barrier* to the total potential freedoms available. This is what makes receiving accurate *perceptions* from *"remote"* *viewpoints* (*ZU-Vision*) difficult for those individuals that continue to *compulsively create* and *unknowingly agree* to the *reality* of the *wall* as a *real barrier* for *Self*.

There are other *agreed-upon barriers* of a *Shared Universe* as well, less obvious perhaps, such as *"space"*—which is also to say *"distance."* And if there is to be any *creation* or *activity* within this *"space,"* then there is also the observable factor of *"time"*—particularly when there is a perceived *motion* across *"distances"* (or even inherently in the "decay" or "erosion" of what is considered *solid-matter*).

For example: *communication* is a relay or motion of a "particle" or "bit" from *"Spot-A"* to *"Spot-B."* It must *"cross a distance"* and thus there is some *"time-lag"* in this action. It would be instantaneous were it not for the perception of some kind of *barrier*, such as *distance*. Without a *spatial distance*, however, there would be no *"Spot-A"* or *"Spot-B"* since the two would be indistinguishable and now essentially the same "spot."

There is a systematic relationship between *communication* and *"proximity"* (or *"closeness"*). This is reflected in the degree of *"likingness"* or *"affinity"* that is shared between ourselves and other individuals and things.

What we *"like,"* we generally want to keep *"closer"* to us; and the more we engage in *true communication* with someone, the more we can get to *liking* them, come into *closer "agreement"* with them (and be even further inclined to *communicate* with them more). We are, of course, only describing a systematic tendency in this case, not an absolute.

But these "factors" of *communication, likingness* and *agreement* are indeed interconnected; and that is how the area of *"barriers"* is initially handled at this *systematic processing level*. What is *"fragmentation,"* but an *energetic-mass*; and what is an *"energetic-mass,"* in an otherwise clear flow or current, but a *"barrier."*

For training and demonstration purposes, it may be helpful to *imagine* these *three factors* as a "current-flow" or "channel" connected between all individuals and things in existence. The "degree," "type," or "intensity" of these *flows* tends to rise or fall collectively. Individuals *communicate*, they *like* each other, and *agree* with each other, *&tc.,* or else they "break," "reject," or otherwise "wall up" against these connections.

It is true that we may initially have "good reason" for establishing *"breaks"* and *"barriers."* However, we also have that interrelationship of factors (*communication, likingness,* and *agreement*) to be concerned with. This means by "cutting" *communication,* one then deals with *dislike* and *disagreement, &tc.* And there is the potential for an *automatic* (or *compulsive*) continuation of a *"break"* or *"barrier" unknowingly.*

More importantly is the fact that when the "event" or "incident" that prompted the *break* is not properly *confronted "As-It-Is,"* then *fragmentation* generally ensues. This means there is a potential for *turbulent emotional* or *mental "charge"* in that entire area thereafter. And this increasingly builds up as *"chain of fragmentation"* connected to other similar "incidents."

For example: an individual that has often had their *affections* (*likingness*) "rejected" is likely to be more emotionally sensitive to that area, or more turbulently reactive at even the most subtle indications of "rejection" in the future. Our systematic solution is to handle the considerations from earlier "rejections" (which is where the primary *fragmentation* or "upset" is actually stemming from).

Breaks and *barriers* (in our three *flow-factors*) can occur from "enforcement" in addition to "rejecting" or "inhibiting" something. "Too much" of something is often just as uncomfortable as its "absence." This includes any time "too much" *force* is applied to any of our *flow-factors*—such as being *"forced to agree."* Quickly, we will shut down *communication flows,* then strongly *dislike* and *disagree* and so on as a cycle.

We say that this type of *fragmentation* occurs as a "chain" that is connected to many incidents—not just the one that *"triggered"* or *"restimulated"* a *reaction.* Usually the *reaction* is also out of "proportion" to what the present incident called for. But in essence, that is not *all* that the individual is *reacting* to; there is an entire "chain" of *fragmentation* accumulated from our past (or "*Backtrack*") that is also now "active" and present.

The systematic solution to these types of "upsets" is to "*spot*" the underlying source of *fragmentation* that has "*resurfaced*" (or is "*in restimulation*"). This actually increases "*Actualized Awareness*" when one is "upset"—as opposed to trying to get someone to "calm down," which is really only a "*suppression*" (and which only further validates and strengthens the *fragmentation* itself).

Systematic Processing in these areas is most effective when the *earliest* incident of a certain type of occurrence can be *spotted*. For a *Seeker's* first cycle through the lessons of the *Professional Course*, only those events taking place during *this* lifetime are *processed* (unless additional data is already readily available); however, in later passes through our *course* material, an advanced application would include "past lives" ("*Backtrack*") too.

DEFRAGMENTING THE "FACTORS"

We have, for our *systematic processing*, three "factors" (or *flow-factors*) that require *defragmentation*: "*communication*," "*likingness*" and "*agreement*." There are also two basic *flow-types* applied to each *factor*: "*insistence*" (or "*enforcement*") and "*rejection*" (or "*inhibition*").

The three *processing command-lines* ("PCL") for each of the following *processes* are *run* in alternation until any sense of "*fragmented energetic charge*" (concerning an area) has fallen away or dispersed. *Run* as many of the *processes* within a single *session* as can be handled—being certain to *end-the-session* when "feeling good" and alert. All of these *processes* employ the "*Analytical Recall*" technique (as first described in *Lesson-2* of this *Professional Course*).

As an additional point of instruction: if any of the *processes* stirs up significant *energetic turbulence,* a more complete *defragmentation* will only occur if an *earlier* instance of a similar type is *spotted.* This may, in fact, have to continue "down a chain" of *even earlier* instances until the *earliest* one available for *recall* is *spotted.* Only then can the actual *fragmented imprint* (underlying the "upset") be handled and *confronted.* Also be sure to *notice* some "things" and "actions" within the incident, rather than simply listing it off as *recalled.*

Communication Factor: Enforced Out-Flows

1. *"Recall a time when you insisted that someone communicate with someone (or something)."*

2. *"Recall a time when someone insisted that you communicate with someone (or something)."*

3. *"Recall a time when someone insisted others communicate with someone (or something)."*

Communication Factor: Enforced In-Flows

1. *"Recall a time when you insisted that someone communicate with you."*

2. *"Recall a time when someone insisted that you communicate with them."*

3. *"Recall a time when someone insisted that others communicate with them."*

Communication Factor: Inhibited Out-Flows

1. *"Recall a time when you insisted that someone not communicate with someone (or something)."*

2. *"Recall a time when someone insisted that you not communicate with something (or someone)."*

3. *"Recall a time when someone insisted that others not communicate with someone (or something)."*

Communication Factor: Inhibited In-Flows

1. *"Recall a time when you rejected someone's communication."*

2. *"Recall a time when someone rejected your communication."*

3. *"Recall a time when someone rejected another's communication."*

Communication Factor: Clear-Flow

1. *"Recall a time when you communicated well with someone."*

2. *"Recall a time when someone communicated well with you."*

3. *"Recall a time when someone communicated well with others."*

Likingness Factor: Enforced Out-Flows

1. *"Recall a time when you insisted that someone like something (or someone)."*

2. *"Recall a time when someone insisted that you like something (or someone)."*

3. *"Recall a time when someone insisted others like something (or someone)."*

Likingness Factor: Enforced In-Flows

1. *"Recall a time when you insisted that someone like you."*

2. *"Recall a time when someone insisted that you like them."*

3. *"Recall a time when someone insisted that others like them."*

Likingness Factor: Inhibited Out-Flows

1. *"Recall a time when you insisted that someone dislike something (or someone)."*

2. *"Recall a time when someone insisted that you dislike something (or someone)."*

3. *"Recall a time when someone insisted that others dislike something (or someone)."*

Likingness Factor: Inhibited In-Flows

1. *"Recall a time when you rejected someone's affection (or attention)."*

2. *"Recall a time when someone rejected your affection (or attention)."*

3. *"Recall a time when someone rejected another's affection (or attention)."*

Likingness Factor: Clear-Flow

1. *"Recall a time when you liked someone."*

2. *"Recall a time when someone liked you."*

3. *"Recall a time when someone liked another."*

Agreement Factor: Enforced Out-Flows

1. *"Recall a time when you insisted that someone agree with something (or someone)."*

2. *"Recall a time when someone insisted that you agree with something (or someone)."*

3. *"Recall a time when someone insisted others agree with something (or someone)."*

Agreement Factor: Enforced In-Flows

1. *"Recall a time when you insisted that someone agree with you."*

2. *"Recall a time when someone insisted that you agree with them."*

3. *"Recall a time when someone insisted that others agree with them."*

Agreement Factor: Inhibited Out-Flows

1. *"Recall a time when you insisted that someone disagree with something (or someone)."*

2. *"Recall a time when someone insisted that you disagree with something (or someone)."*

3. *"Recall a time when someone insisted others disagree with something (or someone)."*

Agreement Factor: Inhibited In-Flows

1. *"Recall a time when you rejected someone's reality (or refused to agree with them)."*

2. *"Recall a time when someone rejected your reality (or refused to agree with you)."*

3. *"Recall a time when someone rejected another's reality (or refused to agree with them)."*

Agreement Factor: Clear-Flow

1. *"Recall a time when you agreed with someone."*

2. *"Recall a time when someone agreed with you."*

3. *"Recall a time when someone agreed with another."*

HANDLING THE "FLOW-FACTORS"

Very early on the *"Backtrack,"* we approached existence from a much more *"All-Pervading"* and *"All-Knowing"* state—but, of course, that did not offer us much room to experience any kind of *game-conditions*. Therefore, to have some genuine sense of *interest* or *curiosity* in our lives, it was necessary for us to first *agree* to *"Not-Know"* at least *something* about something.

Where it comes to our encounters in this present *Shared-Game Universe*, the *"Not-Knowing"* is what allows the original *barriers* to exist. The *agreed-upon* boundaries defined (for example) by an individual's own *"Mind"* or a *"Wall"* restrict the *apparent* "ALL-ness" that could be potentially experienced or *Known*. These *barriers* only exist, of course, within the *"reality"* of the *Game* or *Universe*; they are not *"actual"* conditions.

Most of the time, the "upsets" and "imbalances" that affect us in our daily lives come from the *perception* that there are too many *factor-breaks* and *barriers*. An individual can also become "inhibited," "antagonistic" or "bored" by not having enough *randomness* too. Therefore, a happy healthy life is comprised of just enough *"game"* for one's own *tolerance*. Of course, increasing that *tolerance* is quite desirable for *Ascension*.

We often approach the *Game of Life*, then, from some degree of *"Not-Knowing."* Our *attention* is then directed by *interest* or *curiosity*. This generates an *energy-flow*, such as we have described in our lessons concerning the *factors*, *flow-types* and *circuits*. When there is "good" *communication* and *agreement* (*&tc.*), life plays out with minimal disruption and difficulty.

But, *this Universe* is quite obviously not built upon *Self-Honesty*— and *fragmentation* runs rampant in the typical *"Human"* experience. Often times, an individual does not fully *"connect"* and instead encounters some degree of *resistance*. Assuming one continues to be *interested* in spite of this, an increase or amplification (intensity) of the *energy-flow* (or *attention-flow*) is necessary to overcome the *resistance*. This is where the individual tends to find some trouble.

As we have spoken of previously in this lesson, when an individual starts to apply *energy* against *energy*, and *force* against *force*, they begin to engage with further *barriers* and more solid *game-conditions*. In the example just given: if an *interest* or *curiosity* (in someone or something) is *inhibited*, it promotes an *increased desire* and a more impactful (or solid) energetic or material *effort* to overcome the *resistance*. This brings us far and below the operating levels of high-power *Alpha-Thought* and true *Self-determinism*.

Continuing our example: if the individual is unsuccessful in accomplishing what they *desire*, and is unable (or unwilling) to abandon the pursuit, the directed energy (of the flow) must increase further in its material solidity, and *effort* will now be applied to *enforce* or *insist upon* the intended *flow*. This is when an individual falls low enough in *Awareness* to start operating *automatically* on *reactive-mechanisms* (*fragmentation*).

Once an individual starts *reactively* operating on *fragmentation*, their *efforts* generally are unsuccessful—quite frankly because they are intensely applying *effort* in the first place. Assuming this fails to deliver the *desired results* (or *effect*), as it usually does, the individual switches to the *inhibition* side of the *flow*. The "thing" or "person" (*terminal*) is now so "*highly charged*" that the *intentional effort* becomes "to get away" from it (or to "keep it away"). But, the individual is still *intensely* and *compulsively* "connected" to it.

Rather than confronting a thing "*As-It-Is*," the typical *reactive-response* to "getting away from" something that is *highly charged* (or undesirable) is to "make nothing" of *it*—to treat it as if it were "not a thing." It is no longer experienced "*As-It-Is*," and yet there is still a "*flow*" that an individual now starts rejecting furiously. This is the point when one angrily "doesn't want" anything to do with the thing they formerly *desired*. And in low-level cases, the individual will go as far as to act out against (or even destroy) it.

This systematic sequence we have described may apply to any of the three *flow-factors*. This is the theory demonstrated in the *processes* found in the previous section—which provides a systematic means of sorting out the originating source of *turbulence* and *fragmentation* for such *flows* encountered in everyday life.

When the *Alpha-Spirit* decides to "*Not-Know*" something in order to have a "game"—an application of *attention* and *interest* (or *curiosity*) is what essentially provides something to "do" ("be" or "have") in existence. It provides enough *randomness* for one to enjoy their life. However, when this is not handled properly—or when the "player" is not operating clearly in *Self-Honesty*—succumbing to a level of *intense desire* only imposes further *fragmentation* and *barriers*. Life suddenly becomes more difficult to manage.

The "upsets" of life are treated as *breaks* (or *barriers*) in the *flow-factors*. For *Solo-Processing*, a *Seeker* first looks over the *incident* or *occurrence* carefully, *spotting* and *confronting* whatever is accessible. To make the *fragmentation* available for *processing*, the first step is simply seeing "*What-Is*," and not focusing on any *confusions* or other unresolved parts of the experience.

These *flow-factor* "upsets" are only *systematically processed* when a *Seeker* is distanced from the source of the *break*—when the *turbulence* is not actively *restimulated* by the environment. If a *Seeker* is still too *emotionally charged-up* (hysterically upset, *&tc.*) from a recent *occurrence*, then the first step must include alternating "*spotting something in the incident*" and "*spotting something in the (room)*" until further *processing* can begin.

When the *Seeker* is ready for *defragmentation* of the *upset*, the instructions are: consider each of the *flow-factors*—"*communication*," "*likingness*" and "*agreement*"—and determine which was the most significantly present in the *upset*. Once you have spotted the *factor*, consider what *flow-type* is attached to it; primarily, is it being *"en-forced"* or *"inhibited"*?

Familiarity with PCL from the previous section should help you to identify the "quality" of the particular *flow* you are handling for this form of *processing*. Note that in this lesson, we have also added *"Not-Knowing," "interest/curiosity"* and *"desire"* to this "scale" of potential *flow-types*. These other three types are sequentially "above" (or precede) *"enforcement," "inhibition"* and *"refusal"* on the "scale."

When the correct combination of *factor* and *flow-type* is *spotted* for an *upset*, there should be some sense of *relief* or *emotional release*. If not, it is possible that either the *factor* or *flow-type* (or both) was assessed wrongly. In that case, you simply return to the beginning of these steps and try again. If you "stir" up too much *turbulence* in trying to find out, simply add the alternating PCL of *"spotting something in the room,"* so that not all of your attention is fixed on the *restimulation* of the *upset*.

The *relief/release* gained by this *spotting* technique (above) may be partial or complete. If it's complete, then you can move off onto another *process*, or *end-session*. If, however, it is only partially *de-fragmented* (but there is some *relief* from *spotting* the correct *factor* and *flow-type*), then you continue with additional steps (below).

First, *spot* the *"flow-direction"* or *"circuit."* For example: did *you inhibit* someone's *communication*, or did someone else *inhibit yours?* Then, *spot* exactly what *"communication"* was *inhibited* (for example); and then *spot* what you *"did"* (*action*) and what you *"decided"* (*thought*) in the incident. If possible, also *spot* any lingering upper-level *considerations* and *postulates* you made as a result of the experience.

If this technique doesn't provide a complete *defragmentation*, it is likely that there is an *earlier incident* of a similar nature that is connected to it "on a chain." All you need to do is *spot* the *earlier incident* and *run* through the steps again.

If at any time during this *processing*, things seem to "feel better" and then suddenly the *fragmentation* seems "more solid" again, you likely *ran* the *process* too long. Simply alternate: *spotting* the moment you had experienced the *end-point*, and *spotting* something in the room, until you get to "feeling better" again.

MORE ON "BARRIERS"

Let us now take a moment and examine a holistic view, reviewing what we have *realized*, from all the lessons composing our *Professional Course* for the "*Pathway to Ascension*" so far.

In the beginning, an *Alpha-Spirit* goes "*out-of-communication*" *knowingly* and *selectively*. While this, at first, is a matter of personal preference or choice, it is generally encouraged or influenced from an "outside" or "other-determined" source. This is undesirable, because once an individual goes "*out-of-communication*" too far, they then are suddenly easier to *control* from outside "environmental" and "other-determined" sources.

Once a being is "*out-of-communication*" there is a greater chance for encountering things that they don't *like*; and the individual begins *protesting* them, instead of *confronting* them. This is when *fragmentation* sets in; the individual starts to *compulsively create* (and "*postulate*") things into existence from a state of *protest* (as in "*communicating*" a *protest*)—and this leads to the experience of *problems*.

In typical cases of *fragmentation* (*e.g. the Human Condition*), an individual experiencing "*problems*" will apply increasingly low-level *efforts* to "solve" the *problem*. This leads to committing "*Harmful-Acts*," which now have to be *held-out* (and other actions that are now consciously *held-back*), leading to only further and further *withdrawal* from existence "*As-It-Is*" and the development of additional *barriers*, such as we have described throughout this lesson.

More *recently* on the *Backtrack,* an *Alpha-Spirit* identifies *Self* more closely as a "material body" that can be hurt (because *Self* "*considers*" that *Self* can be hurt as a result). Prior to this, earlier in the *game,* the *Alpha-Spirits* were more like "*demi-gods*" that could really only annoy or tease one another—or mess up each others *creations.* But while *knowingly* operating as an "*eternal being,*" they knew that nothing permanent could affect each other. Of course, we have fallen quite far in our *Awareness* of this native state, and have found ourselves essentially "stuck" within these *creations.*

From the perspective of the *Human Condition,* it often *appears* as though we have "good reason" for the various *upsets* we encounter in life—especially since our recent history on the *Backtrack* has included great "pain" and "destruction." It *appears* as though the *breaks* and *barriers* (of *factors* and *flows*) are a direct result from the various harms that have been present in this lifetime and in recorded history.

The truth is that these *breaks* and *barriers* are connected to "chains" of *fragmentation* that occurred much earlier in our *Spiritual Timeline* (or *Backtrack*); back when we were still "above" the level of *considering* ourselves able to be actually harmed (in a mortal sense). We could, however, still feel "hurt" from, for example, "*refusal to communicate*" or "*rejected affections,*" *&tc.* That is how a lot of the "mess" in existence came into being: first the *breaks* and *barriers,* and then the "*wars.*"

At first, a *Seeker* may question the *systematic accuracy* of the "*fragmentation pattern*" we have just described. After all, personal tastes ("*likes*") often differ, and there are many people presently not "*communicating*" with each other—and this does not seem to present a *problem* or *upset* in itself. But, that is not what we mean in this lesson. Simply not having a lot of *flow*-activity is not the same as having a *factor-break,* in that it does not carry the "*emotional charge*" or a "*chain of fragmentation*" with it.

The *factor-breaks* we experience in this lifetime as *upsets* only occur because there is already existing *"charge"* or *"fragmentation"* present on the line. A proper or "clear" use of the *factors* is usually sufficient to "dissipate" or "eliminate" most of what accumulates in everyday life. For example: when we "listen to," "care for," "help," or otherwise engage well with others and our environment, our lives bring "happiness."

On the other hand, when the *factor-flows* are "cut" or "broken" abruptly—either by *enforcement* or *inhibition*—any of the "charge" that would be "relieved" by a proper *flow*, suddenly "backs up" (or becomes a "blockage") as an *energetic-mass*. This is what prompts a personal response, originating from the *reactive-mechanisms* of *fragmentation*, that seem so excessively out of proportion to what a present situation calls for.

As we close this lesson, it is important to understand: it is not the *factor-flows* themselves that cause *fragmented charge* on one's track, but an individual's own *compulsions* and *inhibitions* in regards to the *breaks* and *barriers* encountered. And this includes the *fragmentation* encountered concerning what an individual "must" or "must not" *Be, do* or *have*. And this returns us full circle to what we have covered in the earlier lessons regarding *"reach"* and *"withdrawal."*

Finally, it is only when we are in a state of *Self-Honesty*, with a *willingness* to *Be, do,* or *have* "anything" (without any *compulsion* or *reactive avoidance*), that we are truly *free* of the *Human Condition*—having risen far beyond the *barriers* of *upsets* and *problems* inherent in that *fragmented* state.

It is only when we can regain the high-power *Alpha-Thought* of our former native state that we will have a truly *Self-determined* and totally *free choice*. In a *fragmented state*, all *considerations* are of a "lower order" because the external "other-determined" factors of life are given more validation as a *source* of our experience of existence than we give ourselves. This is one of the things that improves for a *Seeker* as they progress in their *processing levels* of *Systemology* and prepare *Self* to again experience an *Ascended* state.

ADVANCED PROCESSING

The following *systematic processes* are traditionally applied after using the PCL given in *Lesson-6*: *"Spot three places you are not."* The purpose is to more easily assume a *viewpoint* that is *"exterior"* to the *body* (or even the confines of the *Physical Universe*) and be able to *"spot"* things from that *viewpoint.*

For these *processes*, simply *imagine* that you are hovering above, and freely able to move about, a city or populated area. At this *processing level*, we are not concerned with how accurate or vivid these *"remote" perceptions* (or *"ZU-Vision"*) may be. Just *"spot"* things anyway; *imagining* them (or how they *might* be) as needed.

Attacking

1. *"Spot three people that you are not attacking."*

2. *"Spot three people that are not attacking you."*

3. *"Spot three people that are not attacking others."*

Hatred

1. *"Spot three people that you do not hate."*

2. *"Spot three people that do not hate you."*

3. *"Spot three people that don't hate each other."*

Ordering

1. *"Spot three people that you are not giving orders to."*

2. *"Spot three people that are not giving you orders."*

3. *"Spot three people that aren't giving orders to others."*

Beauty

1. *"Spot three things you find beautiful to look at."*

2. *"Spot three things someone else would find beautiful to look at."*

3. *"Spot three people looking at beautiful things."*

Safety

1. *"Spot three places where your body would be safe."*

2. *"Spot three places where someone else would be safe."*

3. *"Spot three places where other bodies would be safe."*

0. *"Spot three places where you would be safe."*

LESSON EIGHT
"CONQUEST OF ILLUSION"

PERSONAL INTEGRITY

Our practice of *Systemology* toward *Ascension* is effective for the simple fact that: at the core, in our most basic state, the *Alpha-Spirit* is actually quite "righteous" and "good." This is what allows us to rehabilitate *abilities* and restore the level of *Awareness*. Nothing is being "added" to *Self* on the *Pathway*; it is the "additives" that are being "removed," and in doing so, removing the *barriers* and *blockages* of our true state.

We have already treated the preliminaries of *"Eliminating Barriers"* in our previous lesson. In this one, we continue further in this area to complete the necessary *systematic processing* for *Systemology Level-3*. Much of the material found in this lesson is based on a book found in our *Systemology Core*, titled: *"The Way of the Wizard."*

At the inception of *Shared-Game Universes*, *Alpha-Spirits* went to great lengths to "trick" and "deceive" one another. This is because, during our early experiences as *god-like* beings—before we considered our *Self* to be attached to any kind of *"body"* that could be harmed—"deception" and "trickery" is about all that you could really do against another *god-like* being.

"Illusions" were originally created for fun and entertainment—similar to going to the "movies" or attending a "stage magic" show. Remember that, at basic, we as *Alpha-Spirits*, have a uniquely strong fondness for *"games."* But later on, more recently on the *Backtrack* of our *Spiritual Existence*, things became more serious as these "illusions" started to be used as a means of *entrapping* and *enslaving* each other.

This is not an easy area for many to handle or *process*. As such, many *Seekers* that have started upon the *Pathway* often fall by the wayside, never progressing beyond this *processing level*. It is a lot to *confront* for some *Seekers*, which is why we only earn success with these areas after first completing the work provided in the previous lessons of this *Professional Course*.

"CONFUSION" AND "FALSEHOOD"

It is sometimes quite challenging to differentiate between *truth* and *falsehood*, or to effectively *dispel illusion*. We have all had a lot of experience in mastering these skills. *Trickery* and *illusion* is also used to *conceal* the true native state that is available to us as *Alpha-Spirits*.

The *fragmentation* that stands as a *barrier* to the highest *realizations* of ourselves is almost entirely composed of *illusion* and *falsehood*. Our experience on the *Backtrack*, in this lifetime and others, has afforded us a great bit of *fragmentation* concerning all sides of this area—what has been done to us and what we have done to others —and also what we have observed with others handling others.

Some of what would appear to be *"basic processing"* in this lesson is really intended to simply get a *Seeker* to actually recognize the facets associated with these various areas that have contributed to holding them back from the full *realization* of *Self*, in *Self-Honesty*, free of *fragmentation*, *compulsion* and *inhibition*.

When an individual *"Knows,"* they are not susceptible to *"falsehood."* Only after an *Alpha-Spirit* agrees to *"Not-Know"* (in order to have a *game*) will they fall prey to *trickery* and *deception*. One of the ways in which an *Alpha-Spirit* might be fooled, is through *misdirection* of *attention*—which is to say *"distraction."* It is this mechanism that allows a *"stage magician"* to fool an audience— and concealing what is *really* going on.

There is also the matter of *"confusion."* When one is in a state of *"confusion,"* they are quite likely to reinforce the solidity of *fragmentation* in their lives by reaching for the wrong piece of "data" in which to stabilize or orient themselves. States of *"confusion"* are so undesirable that an individual will grab a hold of anything that might bring resolution to the situation without fully analyzing and reviewing the content. And this is one way in which a person might be misled or manipulated to agree with *"false data."*

One of the most basic ways an individual overcomes *illusion* and *falsehood* is by *recognizing* it *"As-It-Is."* This becomes increasingly simpler to do as a *Seeker* elevates their level of *Actualized Awareness*. At high levels of personal application, *Awareness* is sufficient enough to unravel and disperse *fragmentation* "on sight"—but only if *actually* "seen" As-It-Is, and not some false alteration of whatever *"It"* is.

The following *basic processes* employ the *"Analytical Recall"* technique described in former lessons. Their function is to help *realize* the type of areas we want to *resurface*, or bring into view, for *Systemology Level-3*.

Falsehood—Trickery/Fighting

1. *"Recall a time when you tricked someone into fighting."*
2. *"Recall a time when someone tricked you into fighting."*
3. *"Recall a time when someone tricked others into fighting."*

Falsehood—Distraction/Attention

1. *"Recall a time when you intentionally shifted someone's attention."*
2. *"Recall a time when someone intentionally shifted your attention."*
3. *"Recall a time when someone intentionally shifted another's attention."*

Falsehood—Confusion

1. *"Recall a time when you confused someone."*
2. *"Recall a time when someone confused you."*
3. *"Recall a time when someone confused others."*

As an additional exercise in identifying the *ideas* that we have used to resolve or reduce *confusion*, consider the following *process*:

A1. *"Recall a confusion."*

A2. *"What 'idea' resolved or reduced that confusion?"*

DEFRAGMENTING "FALSEHOOD"

There are always people in our societies that "instigate" or propagate *confusion*, *falsehood* and *suppression*. This is not unique to only life here on this planet. There are innumerable times in history when *conflict* or *confusion* was encouraged for someone's personal gain or to eliminate an opponent. In many ways this is an inherent part of *games* that involve *"players"* who perceive themselves as in "competition" with each other.

Information is easily mishandled—or *miscommunicated*. Whether intentionally given or otherwise, such *misinformation* or *false-data* is inherently *fragmentation*. It contributes to our having a false *perception* of our environment, and subsequently our making inaccurate evaluations when participating—*acting* and *doing* things—in existence. This *fragmentation* reduces our *Awareness* and the clarity of our memory.

One of the ways in which *falsehood* is spread, is by *"Passing the Buck."* We mean when *"blame"* is misplaced onto some individual, family, group, or society—race, country, *&tc*. This is a *misdirection* of *attention* (*Awareness*) from a "true source" to a "false source" (thereby disguising the *real* origination of an action).

Another commonly used tactic is to *misdirect* or *"shift"* the data about the date/timing of an event—replacing actual data with invented data. This not only obscures the facts, but can also either *"diminish"* or *"exaggerate"* the *significance* of the data—such as representing *events* as being "longer ago" than they actually are, *&tc*.

Sometimes efforts to *"shift significance"* of facts are more direct. For example, someone might overtly (visibly or obviously) "downplay" or "exaggerate" the perceived importance (*significance*) of some data for their own covert (hidden) purposes or gains. This also occurs when someone acts to *embarrass, belittle* or *shame* an individual. Such *"suppressive"* individuals tend to hinder expressions of *creativity* and *ability*.

The following *processes* employ *"Analytical Recall"* to assist a *Seeker* in *recognizing* and *identifying* the presence of *"misdirection"* and *"manipulation"* in their own lives (and in memory, or on the *Backtrack*). *Processes* like these *may* lead to some new *end-realization*; but more often, directly *confronting* the *imprinted* memory of such areas of *fragmentation* will result in *relief/release* and increased *Awareness*.

Falsehood — Shifting Blame

1. *"Recall a time when you tried to shift blame onto another."*

2. *"Recall a time when someone tried to shift blame onto you."*

3. *"Recall a time when someone tried to shift blame onto another."*

Falsehood — Shifting Time

1. *"Recall misleading someone about the time when something occurred."*

2. *"Recall someone misleading you about the time when something occurred."*

3. *"Recall someone misleading others about the time when something occurred."*

Falsehood — Shifting Significance

1A. *"Recall a time when you exaggerated the importance of something."*

1B. *"Recall a time when you downplayed the importance of something."*

2A. *"Recall a time when someone else exaggerated the importance of something."*

2B. *"Recall a time when someone else downplayed the importance of something."*

Falsehood — Embarrassment/Shame

1. *"Recall a time when you made another embarrassed."*

2. *"Recall a time when someone made you embarrassed."*

3. *"Recall a time when someone made another embarrassed."*

Falsehood — Accusations

1. *"Recall a time when you falsely accused someone."*

2. *"Recall a time when someone falsely accused you."*

3. *"Recall a time when someone falsely accused another (or others)."*

Falsehood — Encouraged Conflict

A. *"Spot a time when you were told that someone (or something) was bad."*

B. *"Identify the person that told you that."*

C. *"Did that person have personal interests invested? How so?"*

Falsehood — True/False

A. *"Spot something you were told that you found out to be true."*

B. *"Spot something you were told that you found out to be false."*

Falsehood — Manipulation

1. *"How have you misled (or manipulated) another?"*

2. *"How has another misled (or manipulated) you?"*

3. *"How has another misled (or manipulated) others?"*

HANDLING "SUPPRESSION"

As a *Seeker* progresses along the *Pathway*, more and more of the "hiccups" and "upsets" of the *Human Condition* — and experience of life in general — are able to be *confronted "As-It-Is"* (how things actually are).

Most of us have encountered at least one individual in our present lifetime that continually "acts" toward *creating* challenges and difficulties for you personally. Some individuals apply this *effort* to larger groups of people, or even society as a whole. Those maintaining "higher" levels of *Actualized Awareness* are generally in a better position to handle such people directly, or at least the situations that they are *creating*.

A highly *Self-Actualized* individual doesn't succumb to suppression. It is possible, sometimes, to "win an enemy over" (getting them to accept/agree with your "side" or *reality*). If you can maintain your own "integrity" or "position" strongly or long enough, sometimes it is sufficient to "drive them off," or simply "wait them out" until the matter is dropped. But, this all requires being free of *automatic reactive-responses*.

There are also those unexpected or intense situations when an individual quickly becomes *overwhelmed*—the sense that one has become *too much* the *effect* of the environment (or external source). The best course of action in this case is to remove yourself from the overwhelming influence and "catch your breath" in a safe position. This isn't a *withdrawal*, when ultimately, the situation will still have to be *confronted "As-It-Is."*

One reason why these persons and situations are so challenging for many individuals to handle, is because, quite often: the "present occurrence" *restimulates* much older (and longer) *"chains of fragmentation"*—stirring up *turbulence* from similar situations in the past that were left unresolved (*not-confronted*). This is what influences our succumbing to external "pressures."

Enslavement and *Domination* are *games* as old as the *Backtrack* itself. In fact, there are some individuals that carry *"fragmented purposes"* with them that are left over from *old games* (often from "past lives") they still have a lot of their *Awareness* "stuck" on. They are still "playing" or "acting" in *this life* and *this game*, the *"roles"* or *"goals"* that they clung to long ago. That creates difficulties for others also playing on *this field*.

While it is easy to identify the more obvious situations we have encountered, it is the *"covert attacks"* and *"passive-aggressive"* efforts to *dominate* or *invalidate* us, that many *Seekers* are unprepared to handle in life. The most dangerous of our supposed "opponents" are those that specifically attempt to *"suppress"* our progress, gains, creative ability, and of course, advancement on the *Pathway-to-Ascension*.

A *personality* or *player* that is a *suppressive-type* will consistently try to *invalidate*—or "make less of"—you. Their general aim for *control* is to *"stop action."* Encounters with such types tend to leave us feeling "depressed" or "hopeless"—and yet they may also attempt to make us, in some way, "dependent" on them.

It is senseless to try to avoid all encounters with *domination* and *competition* in this *Universe*. It is an inherent part of the *"roles"* and *"goals"* perceived as part of the *game*; but many players are perceiving and operating with *"fragmented purposes"*—and they do not improve our experience of the *Game of Life* in any way.

The most desirable position (or solution) is to simply be in such "good condition" (as an *Actualized Seeker*) that these "external" (*other-determined*) *efforts* do not have enough "impact" to *invalidate* you—that your level of *Awareness* "exceeds" the degree or level of *energy* or *force* being "thrown at" you, and it may be essentially dissolved *"As-It-Is."* There is no reason to react or engage with further *energy* or *force*, which are of a "lower" order or condition of *existence* and *creation* than a *Self-Actualized Alpha-Spirit* is capable of.

DEFRAGMENTING "SUPPRESSION"

As mentioned above, most *suppressive control* is aimed at *"stopping motion"* (or action). While there are ways of being *constructively critical*, most *criticism* is intended to reduce or "make nothing" of another's *efforts* or *creations*. The following *basic processes* will help in identifying these *occurrences*. [We will employ the word *"spot"* in these PCL to more accurately apply this *processing* to the full *Backtrack*.]

Suppression—Stops

1. *"Spot a time when you stopped someone."*

2. *"Spot a time when someone stopped you."*

3. *"Spot a time when someone stopped another (or others)."*

0. *"Spot a time when you stopped yourself."*

Suppression—Criticism

1. *"Spot a time when you criticized someone."*

2. *"Spot a time when someone criticized you."*

3. *"Spot a time when someone criticized another (or others)."*

0. *"Spot a time when you criticized yourself."*

Suppression—Invalidation

1. *"Spot a time when you 'made nothing' of someone."*

2. *"Spot a time when someone 'made nothing' of you."*

3. *"Spot a time when someone 'made nothing' of another (or others)."*

0. *"Spot a time when you 'made nothing' of yourself."*

It is not uncommon to find that encounters with a very *"suppress-ive-type"* during childhood can remain influential later on in life— even when that individual is no longer present. But that is how *imprints* work. When *Solo-Processing* the following exercise, use your *notebook* or *Flight-log* to record the data from each PCL-question. List multiple answers (if necessary) to make sure you have answered each fully.

A. *"Is there anyone around whom you seem to become 'sick' shortly after seeing them?"*

B. *"Is there anyone who is continuously critical of you?"*

C. *"Is there anyone who is often telling you how 'bad' other people are?"*

D. *"Is there anyone who continuously 'stops' you?"*

E. *"Is there anyone who continuously 'invalidates' you?"*

F. *"Is there anyone who often provides 'false' information?"*

G. *"Is there anyone who often 'makes nothing' of your efforts?"*

If you find yourself writing the same name on several lists, then *run* the following *processes* using that name as the *"terminal"* (blank space) in the PCL. If there are multiple names that frequently appear, then *run* those names as *terminals* too (each as their own series of *processes*). There is no "judgment" applied here about someone else's *intentions* toward us. The very fact that their name shows up frequently on these "lists" is sufficient to indicate there is enough *charge* present (as *fragmentation*) to be *run* as *processing*.

The following series of *processes* begins with a familiar one (from previous lessons) regarding *"help."* The purpose is not to enforce actually helping someone, but to restore the free choice (that is not overridden by *fragmentation*). *Considerations* of *"help"* also assist in breaking down the *barriers* of negative reactive emotion so that a *Seeker* is better able to actually confront the source of *suppression* *"As-It-Is."*

Suppressive Terminal—Help

1. *"How could you help ---?"*

2. *"How could --- help you?"*

3A. *"How could --- help others?"*

3B. *"How could others help ---?"*

Suppressive Terminal—Problems

1A. *"What 'problem' have you been to ---?"*

1B. *"What have they done about that?"*

2A. *"What 'problem' has --- been to you?"*

2B. *"What have you done about that?"*

3A. *"What 'problem' has --- been to others?"*

3B. *"What have others done about that?"*

4A. *"What 'problem' have others been to ---?"*

4B. *"What have they done about that?"*

Suppressive Terminal—Hold-Outs

1. *"What haven't you said to ---?"*

2. *"What hasn't --- said to you?"*

3. *"What hasn't --- said to others?"*

Suppressive Terminal—Actions

1. *"What have you done to ---?"*

2. *"What has --- done to you?"*

3A. *"What has --- done to others?"*

3B. *"What have others done to ---?"*

Suppressive Terminal—Invalidation

1. *"How have you 'invalidated' ---?"*

2. *"How has --- 'invalidated' you?"*

3A. *"How has --- 'invalidated' others?"*

3B. *"How have others 'invalidated' ---?"*

Each of the next three *processes* contains six alternating PCL. They are *run* using *"Analytical Recall"* techniques—although in this case, there may sometimes not be any available answer for a certain question. Any *turbulence* or *resistance* attached may then be handled using what you learned about *factors* in *Lesson-7*.

Trouble Sources—Likingness Factor

1A. *"Is there a time when you rejected their affection (or attention)?"*

1B. *"Is there a time when they rejected your affection (or attention)?"*

2A. *"Is there a time when you insisted that they like you?"*

2B. *"Is there a time when they insisted that you like them?"*

3A. *"Is there a time when you did like them?"*

3B. *"Is there a time when they did like you?"*

Trouble Sources—Communication Factor

1A. *"Is there a time when you rejected their communication?"*

1B. *"Is there a time when they rejected your communication?"*

2A. *"Is there a time when you insisted that they listen to you?"*

2B. *"Is there a time when they insisted that you listen to them?"*

3A. *"Is there a time when you communicated well with them?"*

3B. *"Is there a time when they communicated well with you?"*

Trouble Sources—Agreement Factor

1A. *"Is there a time when you refused to agree with them?"*

1B. *"Is there a time when they refused to agree with you?"*

2A. *"Is there a time when you insisted that they agree with you?"*

2B. *"Is there a time when they insisted that you agree with them?"*

3A. *"Is there a time when you did agree with them?"*

3B. *"Is there a time when they did agree with you?"*

Trouble Sources—Handling

A. *"What could you confront about ---?"*

B. *"What action of --- could you be responsible for?"*

C. *"What about --- could you be at cause over?"*

"JUSTIFICATION" & "RESPONSIBILITY"

Although we have discussed *"justification"* and *"responsibility"* briefly in passing up to this point of the *Pathway*, it is here at the end of *Systemology Level-3* that they are handled more directly. Although these words may appear commonly used, for *systematic processing*, we are primarily concerned with automatic reactivity regarding *compulsive action* (or *compulsive reach*) and *avoidance of action* (or *reactive withdrawal*).

The concepts of *"reach"* and *"withdrawal"* are familiar to us (by this point of the *Professional Course*), but to this, we also add the *"willingness to be responsible."* And by this, we don't mean "being blamed for things." We are speaking of a much higher-level of *"responsibility"* —as in: *"willingness* to be *at cause* over things."

Whenever an individual does something that they consider wrong—whether intentionally or otherwise—there is an *automatic* tendency to attempt to *"justify"* these actions. We do this when *communicating* about such things to others (assuming we don't *hold-out* on the information altogether)—but we also communicate these *"justifications"* to ourselves, and those *considerations* continue to affect us in the future.

In systematic terms: *justification* is a mishandling of *imprinting* or *"charge"* accumulated from an event (and possibly connected like a "chain" to similar events occurring earlier on the *Backtrack*). It strengthens the influence of *fragmentation*, because the "experiential data" is being *altered* rather than *confronted "As-It-Is"* — meaning: *reality* is being presented or perceived as *other than* it actually is; the worldview is distorted.

Maintaining (or *compulsively creating*) *"justifications"* or *"falsehoods"* conflicts with the state of *Self-Honesty* sought as a means of *Ascension*. Maintaining any kind of *"false data"* or *fragmented thought* also leads an individual toward *fragmented actions* that are then used to support (or prove) the *justifications*.

The following *processes* are best *run* as "listing exercises," so be sure to have your *notebook* or *Flight-log* handy. They assist in later bringing a *primary "justification consideration"* into view (which the *Seeker* is operating with the most during this lifetime)—which is the *end-point* of *Systemology Level-3*.

Justification—General
1A. *"What have you done to another?"*
1B. *"How did you justify that?"*
2A. *"What has another done to you?"*
2B. *"How have they justified that?"*

3A. *"What has someone done to others?"*

3B. *"How have they justified that?"*

Justification—Excuses

A1. *"What do you often use as an excuse?"*

A2. *"What do others often use as an excuse?"*

B1. *"How could you survive without excuses?"*

B2. *"How could others survive without excuses?"*

Willingness—Improvement

1. *"What are you willing to improve?"*

2. *"What are you willing to allow others to improve?"*

3A. *"What would someone be willing to have you improve?"*

3B. *"What would someone be willing to allow others to improve?"*

Willingness—Responsibility

A1. *"What could you be responsible for?"*

A2. *"What could another be responsible for?"*

B1. *"What would be acceptable to be irresponsible about?"*

B2. *"What would be acceptable for another to be irresponsible about?"*

[*and then run*]

1. *"Spot a time when you made someone be responsible."*

2. *"Spot a time when someone made you be responsible."*

3. *"Spot a time when someone made another (or others) be responsible."*

[*and finally, run alternately (until the "glee" of "irresponsibility" is discharged)*]

A. *"Get a sense of the joy of responsibility."*

B. *"Get a sense of the joy of irresponsibility."*

DEFRAGMENTING "JUSTIFICATION"

For most of our participation in *Shared-Game Universes*, *Alpha-Spirits* have been operating from a point of *fragmentation*. There is a mistaken, albeit *implanted*, "idea" or "goal" that we must be in *competition* with one another—that *"there can be only one"* or *"only one will survive."* There is, of course, no *actual* truth to this.

In order to reclaim the memory and abilities of our "past," it is necessary to be able to *confront* our actions (and the actions of others) in the "past" without *regret*, without *withdrawing* from the *imprints*, and without resorting to *justification* and *falsehood* (*altering* them) in order to make them more manageable. In *systematic processing*, we take some of the pressure off a *Seeker* by alternating PCL with *spotting* more positive things.

Domination—General

A. *"How have you dominated others?"*

B. *"Spot a way to enhance others."*

Domination—Stopping

A. *"How have you stopped others?"*

B. *"Spot a way to help others excel."*

Domination—Inhibition

A. *"How have you inhibited the survival of others?"*

B. *"Spot a way to aid the survival of others."*

"Domination" and *"Superiority"* are not the only *games* available to *Alpha-Spirits*. There are *better games* at higher levels of existence. However, in *this* present version of a *Physical Universe* (*"Beta-Existence"*)—where physical *energy* and *force* are quite solid and supreme—these *games* are inherently the most common.

The type of "artificial" *fragmentation* that is "*implanted*" as *reality-*

agreements for this *Universe,* instills a certain pattern of "*roles*" and "*goals*" for *this* particular *game.* This means that *all* of us have *something* that we use in order to gain our "*superiority*" over others —and we refer to this as a "*justification consideration.*"

This primary "*justification consideration*" will be some basic characteristic that we emphasize in exclusion to other things—so it *should* be easy to *recognize.* The *consideration* that we are after is also quite basic in structure. It will be a *basic goal,* such as: "to be holy" or "being intelligent" or "to be strong," *&tc.* Whatever it is, it is treated in *processing* as an "item" used to make yourself "*superior*" to others.

It takes this far along the *Pathway*—the amount of *Actualized Awareness* gained by this point of *systematic processing*—in order to *confront* such a *consideration* "*As-It-Is.*" The average *Human* will quickly *withdraw* from handling such things. But even the most "*Enlightened*" among us will still have this "*item*" on their "*track*"—the key difference being that they may have risen above the tendency to use it *against* others.

Although a more complete understanding of *games, goals* and the *Universes* representing them, is generally earned in the most *upper-levels* of *Systemology,* it may be easily stated here that: employing your best characteristics in the low-grade *games* of *force* and *superiority* is quite "*unenlightened.*" It causes a gradual deterioration of *abilities* and abandonment of a *goal* because of the *Harmful-Acts* that are required to maintain that specific type (or "*personality-role*") of "*superiority*" and "*rightness*" for "*dominance.*"

Identifying and *recognizing* this primary "*justification consideration*" is not only necessary for completing *Systemology Level-3,* but is also the critical "*Key*" that "unlocks" (what we refer to as) the "*Backtrack*" for *upper-level processing* and additional "passes" through the *Professional Course* material.

The *game* of "*goals*" and "*roles*" extends back many "lifetimes" and many *Universes.* There is even an observable "systematic pattern" behind the way one *personality-type* and/or *goal* is abandoned, requiring the individual to shift over to another that has not already deteriorated. So this particular "item" we are looking for is likely to be the same for a certain amount of former "lifetime" experiences, but not necessarily many (relatively speaking, compared to the larger scope of things).

For achieving a basic state of "*Beta-Defragmentation*," we primarily emphasize *this* lifetime. But, there are higher applications of this same *Professional Course* after it is completed once—such as additional "*Alpha-Defragmentation*" of the *Backtrack*. For present purposes, however, it is important to prefix certain PCL (such as our "search and discovery" of the primary "*justification consideration*") with "*In this lifetime...*" Later on, a *Seeker* can explore the *Backtrack* further.

It is important to approach this final area of *Systemology Level-3* with care, because accuracy is the only way to effectively gain any stable progress from it. A *Seeker* should be in a state of high *Awareness* (having achieved "good" results from all former *processing*), be well rested (and alert), free of any actively *restimulative* sources of *fragmentation*, and finally, *interested* and *willing* to actually "discover" the "answer."

For this exercise, *list* your "answers" clearly. It's important not to do this while tired, because if you suddenly feel "heavy" or "tired" while *listing*, then you know you have *listed* too many answers—and that the basic answer you're looking for is already on the *list*. If after *listing* for a while, you start to feel "irritated" or "restimulated," then you still haven't put the answer you're looking for on the *list* yet.

Although this is a *listing exercise*, your *list* is not intended to be very long—in fact, you may *realize* what the answer is quite quickly. But, the basic instruction is to *list* "answers" in your *notebook* (or *flight-log*) below where you have written out the "question" (PCL). You continue to *list* until you feel you have "*The Answer*" or you feel comfortable that you have *listed* enough.

When you find *"The Answer,"* this exercise is complete. If you're unsure which it is among your *list*, then go through it and consider each one in turn. When one doesn't seem like it's the right answer, put an "X" next to it, so you don't have to keep considering it. If at any point while checking your *list*, you suddenly *recognize "The Answer,"* then you stop working with the *list*. And, of course, if the *list* seems incomplete, you can always add to your *listed* answers until you've completed the exercise.

The *"Question"* (PCL) is:

"In this lifetime, what makes you superior to others?"

Then, when you have *"The Answer,"* write it in capital letters in your *notebook* (or *log*) and underline it. Take that "item" and use it to complete the next *process. Run* the following three PCL in alternation.

A. *"How does --- make you superior?"*
B. *"How could you use it to make yourself right?"*
C. *"How could you use it to make others wrong?"*

On the other side: we also want to find out what it is about others that you are (or might be) using against them. This might or might not relate to the *"roles"* and *"goals"* attached to the previously discovered *"justification consideration."* To find out, a *Seeker* performs another *listing exercise* (using the same basic instructions as before). This time, the PCL-*question* is:

"In this lifetime, what is it about others that makes them so wrong?"

If you aren't satisfied with the resulting answers, there is another way of using *processing*-PCL for a "search and discovery" (related to the previous exercise) which is:

"In this lifetime, what do you use to make others wrong?"

All of the exercises/*processes* in this section focus on finding out a *Seeker's "justification consideration"* (which is also tied to the *"roles"* and *"goals"* they are playing at in this lifetime). What it generally comes down to is a *basic consideration* that "others are evil," "others are stupid," or "they're weak," *&tc.* [These examples would respectively apply to *goals/roles* of "being good," "being intelligent," and "being strong."]

To *defragment* the *"justification consideration,"* we want to systematically untangle the *"negative"* associations of using our abilities against others, while still retaining the original skill ("intelligence" or "strength," *&tc*).

The following *processes* are not *run* as simple alternating PCL. A *Seeker runs* a single PCL for as long as it continues to produce answers, then shifts to the second to get as many answers for that one, and then shifts back to the first PCL, and so on until the influential "intensity" of the *consideration* is dispersed (falls away).

Justification Consideration — Right/Wrong

A. *"Spot ways that --- would make you right."*

B. *"Spot ways that --- would make others wrong."*

Justification Consideration — Domination

A. *"Spot ways that --- would help you escape domination."*

B. *"Spot ways that --- would help you dominate others."*

Justification Consideration — Survival

A. *"Spot ways that --- would aid your survival."*

B. *"Spot ways that --- would hinder the survival of others."*

The influential "intensity" that we speak of is the sense that the *consideration* is "true." When it is first treated in *processing*, the *consideration* may actually seem very much like the "truth" about "why people are the way they are" — and a *justification* for "why we are the way we are." By *spotting* this *consideration* with high-level *Awareness*, the "*charge*" behind the idea or concept (as being "true") starts to dissipate or disperse. A *Seeker* then *realizes* that this is a falsely held "belief" that keeps them from *Self-Honesty*.

Assisting or "helping" others improve in the same area as your "*justification consideration*" is one way that this *defragmentation* is maintained even after *processing sessions.* It also helps to keep the "*goals*" and "*skills*" themselves from "deteriorating." For example: if your *goal* is "to be strong" (or a *role* of "being strong") and you viewed others as "wrong" for being "weak," the solution is to help others become "stronger," &tc.

This area is a very critical area for rehabilitation of other spiritual abilities and for rising above the limitations inherent in the stand-ard-issue *Human Condition.* Therefore, a *Seeker* should not pass by this "checkpoint" on the *Pathway* without making certain that the "*justification consideration*" is properly found and *defragmented.* Such completes a *Seeker's processing* for *Systemology Level-3.*

LESSON NINE
"CONFRONTING THE PAST"

EXPERIENCING EXISTENCE

This lesson begins with some upper-grade philosophy to lead off *Systemology Level-4*. It is rather important for a *Seeker* to have completed *all* previous lessons of this *Professional Course* (and preferably also the *Basic Course* that precedes it) before continuing onward with the *Pathway* from this point.

Throughout this *course*, a *Seeker* has gradually increased their *willingness* (or *tolerance*) for "facing up to," "looking at," or else *"confronting"* various *"things."* Our systematic approach began with peeling away lighter surface "layers" before progressing into "heavier" (more *"turbulent"*) expressions or manifestations of *fragmentation*. The "logic" behind the "sequence" of procedures used on this *Pathway* becomes more apparent as we start handling "deeper"—more *"repressed"*—layers of *unknowingness* and *fragmentation*.

Properly "handling"—or *confronting*—the "past" is treated (or *processed*) no differently than when handling "present" *existence* (or *conditions* thereof). For purposes of *defragmentation*, there really *is no* difference—because a *fragmented* individual, themselves, is not actually handling a proper "distinction" of *"time."* For them: *"past imprinting"* dictates *"present perceptions"* and *"future action."* *Time* becomes entangled altogether.

The inherent nature of *fragmentation* is that it "sticks" an individual's *attention/Awareness* somewhere on their *"Backtrack"* and then "tricks" them into continuously maintaining a *creation* of that *reality as "present-time"* well afterward. This continuous (and *Self-propagating*) *creation* (and carrying) of *fragmentation* causes us to experience *restimulated emotional turbulence, compulsive reaches, re-*

active withdrawals—and *entrapment* in *barriers* (and other *considerations* about *reality*) we have *agreed* to, even if we have since forgotten.

In previous *processing levels,* we treated some of what we *"hold-out"*—and *"hold-back"*—from expressing toward others; but these types of *"things"* are more readily *known* to us. To proceed further on the *Pathway,* a *Seeker* must also *"uncover"* or *"resurface"* those *"things"* which they even have "hidden" from themselves. So, this is the next area we *systematically* target with *processing.*

CREATION AND EXISTENCE

In *Systemology Level-4,* we restore *Actualized Awareness* and increase a *Seeker's* ability to *confront* (or face a *thing "As-It-Is"*) by bringing things into view that the *Seeker,* themself, has *"repressed."* This is to say: they are *"protesting"* the *existence* of it—attaching *considerations* to "make-*Nothing*-of-it"—even if *unknowingly.*

Various intellectual philosophies have been drawn around a human understanding of *"conditions"* and *"states"* of *Existence.* A continuing student or *Seeker* (from previous lessons) should have little difficulty in understanding and applying our *systematic* approach to this subject; but it is important that this subject *is* understood before attempting to progress further on the *Pathway* (using our *processing* methodology).

We frequently use the phrase *"As-It-Is"* in our *Systemology.* This is not fancy word-play. We mean it exactly "as-it-is" intended. The truest form of a *creation* is the "thing" *As-It-Is; exactly As-It-Is.* The problem with treating "existence," is that an individual is likely to *consider* what it "was" (*past*), and "might be," "could be," or "will be" (in the *future*). Still: none of those *figuring-thoughts* and *considerations* are the "thing" *As-It-Is.*

Fragmentation of *"perception"* and *"consideration"* is what prompts *continuous compulsive creation* of the *Physical Universe* (*Beta-Existence*). *"Perpetuation"* and *"continuation"* are *considerations* of *Time*; and it is only when things are consistently *"altered"* and *"changed"* from their original state that they *"continue"* in *Time*.

At this *level*, we are most concerned with how this philosophy applies to *personal defragmentation processing*; meaning, in regards to an individual's own *creation* (and *experience*) of a *Personal Universe* based on the *"impressions"* received from environmental/external *"cues"* (such as *"forms"* and *"bodies"*) in *Beta-Existence*.

An individual would not be so entrapped by the *Human Condition* or participation in *this Physical Universe* if they experienced *Existence "As-It-Is."* When in a state of *Self-Honesty*, the *fragmentation* and *illusion* of the *Physical Universe* is not nearly as *"solid"* and *"restimulative."* But, of course, this is not the position that most individuals (as *Alpha-Spirits*) are operating from—thus, *fragmentation* perpetuates and becomes more solid.

The experience of *Existence* from the viewpoint of the *Human Condition* is not a *Self-Honest* experience of *Existence "As-It-Is."* Instead, it is an experience based on that same *Existence* "plus" additional *considerations* and *perceptions* specific to the *Human Condition*. That means *"it is altered"* from its original state in order to be experienced at all from that viewpoint. Rather than *"As-It-Is,"* we experience *"It-As-Changed."*

In terms of *systematic processing*, *"fragmentation"* and *"alteration"* (*"It-Is-Altered"* or *"It-As-Changed"*) are essentially synonymous. Entangled and turbulent energy composing *fragmentation* is "suspended in place" (*persists* or is *continuously created*) only because it is an *"alteration"* or *"deviation"* of *"truth"* or *"actuality."*

The *"reality"* that is *agreed-to* and *experienced* is a "lower class" of existence than what things *"actually"* are. A thing *is* what it *is* because it *is*. Everything beyond that is a *consideration* or *associative-thought* related to other experiences, data, or vocabulary, *&tc*.

For example: a particular title of a book may have many copies in existence and is neither rare nor scarce. However, an individual may choose to "hold on to" a certain copy they have, because of some attached "sentimental" value. Perhaps a beloved family member gave it to them. Its *"IS-factor"* has been *"altered."* While the *actual* book might be found in every thrift-store in the city (and could theoretically be given away and then recovered/created at will), the individual feels *"compelled"* (*compulsion*) to retain their *"changed"* copy.

We are not trying to make a point here that a *Seeker* should give up any "sentiment" on all things. We are simply using this example to illustrate why we retain (and *compulsively create*) our *altered* version of *existence* as personal *fragmentation*—and how easily we can *alter* our personal *reality (experience)* of things.

All of our past experiences on the *"Backtrack"* seem so personally unique and significant that we tend to not want to let them go. And most of them do not have a strongly negative effect on us. However, there are also those things that continue to negatively influence our *perceptions*—those things we don't want to "face" or *confront*—which "persist" *unknowingly* because of how far we have *altered* and pushed them "out-of-view."

We have then described *three* primary *"IS-factors"* of *Existence*: there is the true or pure form as originally created or intended *"As-It-Is"*; there are the additions or alterations of *"It-As-Changed"*; and finally, when what is *actually* there is not *confronted*, but is instead *altered* to the point where it is undesirable to "look" at, we *consider "It-As-Nothing"* and try to make *"it not exist"* by means of applying additional *energy* and *force*.

This accurately describes the entire *systematic* "sequence" of *repressed* deeply-laden *fragmentation*. Of course, when *"resurfaced"* in *systematic processing (defragmentation)*, the "sequence" is experienced in reverse. When the *"It-As-Nothing"* (and *"Not-Known"*) *considerations* are initially lifted, the *"It-As-Changed"* (or *"Almost*

Known") and *imprinted considerations* come into view first. These must be *processed* before the true *"As-It-Is"* (or *"Actually Known"*) *creation* may exist or not (at will) with full high-power *"Alpha-Thought."*

[There are many "upper-level" applications of this philosophy (concerning *"Universes"* or *"objective reality"*) that some *Seekers* have had intermittent and sporadic experiences with—which causes them to run astray *before* completing the *Pathway.* Any such *"psychic phenomenon"* (or however you choose to classify it) is really a byproduct of *Beta-Defragmentation* and not a targeted "goal" attached to this *Professional Course.*]

HANDLING EXISTENCE

Fragmentation consisting of *"It-As-Changed"* (*altered*) content is a persistent trouble source that there is at least *some Knowingness* of. But once it is pushed to a level of *"It-As-Nothing"* (*deeply repressed*), it is not only a persistent trouble source, but also entangles potential *Actualized Awareness* in its *continuous creation* "out-of-sight." It still exists. There was *first* an *Altered Something* to *"protest"* the "existence-of" and *then* an attempt to make *Nothing-Of.* But, it is not *actually* "erased" or "destroyed" *As-It-Is* if treated as something other than *As-It-Is.*

For example: a woman receives jewelry from her lover. For our purposes: she has *imprinting* from *past experiences* that *alters* her perception of this activity. Perhaps a former lover was in the habit of giving her jewelry after having an affair. This produces *reactivity* toward the *past* that is now applied to the *present.*

Instead of having *confronted* the *past imprinting*, perceptions of the present situation are being *altered* with emotionally turbulent *"mental imagery"* of cheating lovers, *&tc.* As this *restimulation* becomes more intrusive and harder to *confront*, more effort (*attention* and *Awareness*) is applied to "push it" further "out of view."

During *systematic processing*, once the *repressed* nature of the *fragmentation* can be lifted, the first *"mental imagery"* that comes into view is the "cheating lover." But this data may not consist of the entire *"chain-of-fragmentation."* It is likely that other similar earlier negative experiences (in the past) also contribute to the consistent *repression* of that area of *fragmentation* when left *not-confronted*. This *deep repression* affects our ability to *confront* (and even properly "remember" or "recall") entire areas of our past (*Backtrack*) *As-It-Is*.

The *Alpha-Spirit* is essentially a representative of the *Infinity-of-Nothingness*. As a "balancing factor," its native position (or state) consists of the ability to *create* a *Potential Everythingness*.

At our "highest" *god-like* state, we would want to have *Everything* "available" to us for *creation*—to have the full range of *creating* or *not-creating* things (at will) in *Self-Honesty*. This state requires no *fragmentation* to "unknowingly" or "reactively" influence a *compulsive reach* (*compelled toward*) or *repulsive withdrawal* (*repelled away*) to/from anything. Such *fragmentation* gradually reduces the total *"god-like abilities"* (*Actualized Awareness*) once available to the individual; and *defragmentation* gradually "rehabilitates" these conditions.

We will start off lightly applying this high-level theory and philosophy in practice (for *processing*). For example: an individual might be worried that some undesirable thing might be true, or perhaps something might have been done to them in the past; but in either case, the *uncertainty* produces a *confusion* in that area. This, in itself, generates personal feelings of *"withdrawal"* or *"reluctance to reach"* to some degree.

Now, if this area were *confronted* properly, the *actual truth* would either be *Known*, or at the very least, there would be no reactive reluctance to *"find out."* At higher levels of *Self-Actualization* (*Awareness*), an individual wouldn't necessarily even "care" either way, because it wouldn't persist enough to be troublesome. But, if this can't be achieved *Self-Honestly* by simply "looking at" (or "*confronting*") something *As-It-Is*, then "deeper" *fragmentation* is present, and inhibiting an experience of *clear perception* (*Knowingness*).

There are many techniques described throughout this *Professional Course* that, once understood and mastered, may then be applied as single applications to particular situations. Most of them are designed to increase one's *ability* and *willingness* to "*confront*" something (or an area) with increased *Awareness*. For example: a *Seeker* could *imagine* various "*mental images*" of a thing happening, *copying* them next to each other many times, until they can be "created," "looked at" and "thrown away" easily.

If supporting lessons (background instruction) on this type of approach (or other techniques) are not readily available (or presently known to the student) consider *running* a *process* this wise on something persistently worrisome or troublesome:

1. "*Recall (or imagine) a time when you did something like this to someone (or caused it to happen to someone).*"

2. "*Recall (or imagine) a time when something like this happened to you (or something like this was true).*"

3. "*Recall (or imagine) a time when someone did something like this (or caused it to happen) to another.*"

These *processing command-lines* ("PCL") are alternated repeatedly to "build up" *Awareness* on an area. Using "*recall*" is preferred; but if an actual occurrence cannot be easily "*spotted*" on the *Backtrack*, the *Seeker* can "*imagine*" an appropriate approximation. And this is not to say that the real source of the *basic fragmentation* is not "buried" beneath many layers of, what is likely to be, "*partially imagined data*" anyways. This *process* may be *run* as needed to compliment other related *processes* for a particular area (or occurrence).

BASIC PROCESSES

Each one of the following *processes* may only bring something into view, or it may *defragment* the area altogether (on inspection). PCL are run in alternation. They target the *"It-As-Nothing"* level of *consideration*. When a point is reached where the *process* seems complete, or you "feel good" about it, then *end* it. If something "hidden" came into view, but still feels unresolved, then also *run* the previous *process* given above.

PROCESS #1

1A. *"What might you avoid thinking about?"*

1B. *"What don't you have to avoid thinking about?"*

2A. *"What might someone else avoid thinking about?"*

2B. *"What wouldn't someone else avoid thinking about?"*

PROCESS #2

1A. *"What might you pretend never happened?"*

1B. *"What would be acceptable to have happened?"*

2A. *"What might someone else pretend never happened?"*

2B. *"What would someone else find acceptable to have happened?"*

PROCESS #3

1A. *"What might you have made nothing of?"*

1B. *"What wouldn't you need to make nothing of?"*

2A. *"What might someone else have made nothing of?"*

2B. *"What wouldn't someone else need to make nothing of?"*

SYSTEMATIC DEFRAGMENTATION

At this point of the *Professional Course* (and the *Pathway*), a *Seeker* will have *run* hundreds of *processes* and mastered dozens of techniques for handling various difficulties of *Human Life*. The *processing methods* given in these course lessons are also supplemented with an education of our unique *systemological philosophies*.

As a *Seeker* applies *processing* to different areas of their life, some of them will untangle and unravel—or *defragment*—areas sufficiently enough to "fall away" and cease to affect them altogether. There will be others that are more deeply *fragmented*—producing greater *turbulence*—and the *processes* already given may only "soften" them enough to be more manageable (in *processing* and *everyday life*), but they continue to get "*restimulated*."

It requires increased *Actualized Awareness* (and an ability to *confront*) for one to *actually see* how *Self* is "tricked" into essentially *creating* and *solidifying* its own *entrapment*. It is far easier to pass this *responsibility* or *cause* onto some "other-determined" *Source*—but by doing so, we "mask" or "disguise" (*alter*) the *truth*, and therefore relinquish the *control* necessary (and due to us) to properly handle and manage our *creations*.

The *processing* regimen (and lessons) provided in the *Professional Course* introduces all areas of our *Systemology* with an aim at "*Beta-Defragmentation*"—meaning a handling of *this lifetime*. But, of course, the underlying reasons, mechanics and *considerations* that lead to "*Human Problems*," and troublesome areas of life, stretch back on a "*chain-of-fragmentation*" extending to many *lifetimes* in many *universes*.

For example: a *Seeker* that is able to master the training and *processing* of *Lesson-4* should be able to take apart (and handle) the "anatomy" of *Human Problems* sufficiently enough to experience some relief. But, of course, many of the *earliest* "problems" we experienced on the *Backtrack*—knowingly as "god-like" *Alpha-Spirits*—are still with us, influencing and impinging on the manifestation of *Existence* that we experience everyday.

There are many areas of our life that have *fragmented energetic-*

charge attached to them—entangled personal *energy* and *Awareness* that is "wound up," and just waiting to "spring" (manifest) upon "activation" or *restimulation*. If you have been progressing well on the *Pathway* up to this point, then you have likely experienced some significant *"release points"* and *"end realizations"* along the way. These help increase *certainty* and personal *confidence*, in addition to heightened *Awareness* and ability to *discharge* the *fragmentation* that is *restimulated* by the general experience of everyday life.

Of course, *processing* on the *Pathway* is not "fool-proof." There are times, particularly when *Flying-Solo* (and without additional aids, such as a *"biofeedback device"*) that a *Seeker* may find difficulties or overwhelming *turbulence*. This *Professional Course* was designed to apply to the largest number of potential *Seekers*, thus our emphasis has been on *Solo-Processing* techniques that are effective in *"destimulating" fragmentation*, even in the presence of a "heavy" *charge*.

More often than not, we are intentionally *"resurfacing"* or *"restimulating"* a specific thing in *processing* in order to *"discharge"* (and *defragment*) the *turbulence* (and *fragmented considerations*) directly. We do this *knowingly* and *systematically* to handle with *processing* what would otherwise be *restimulated* by other-determined *sources* and simply build-up as additional *fragmentation*.

To be effective, these *higher processing levels* (that our *Professional Course* now approaches on the *Pathway*) require that a *Seeker* have already reached a certain point of "relief and release" from previous *levels*. It is important to have the *confidence* and *ability* to "cool down" *turbulent upsets* when they are *restimulated* before directing *processing* toward *fragmentation mechanisms* on a more general basis. In brief: you deal with what is already happening, restimulated, or available, before you go digging around deeper for more.

The best course of action in maintaining progress and earning stable results is to practice "preventative" applications of *Systemology*. This means handling the key areas—or *"hot buttons"*—that contribute to accumulating *fragmentation* during everyday *Human* experiences. The *fragmentary dust* of these "normal" *upsets* may be "brushed off" before accumulating into an overwhelming *charge*.

At this time we will identify some *"hot buttons"* from former lessons. As *upper-level* applications, they serve as "preventative fundamentals" to handling everyday life as a more *Self-Actualized* individual—but also to ensure greater effectiveness of *upper-level processing*. Should any of these *"hot buttons"* be actively "pressed" or affecting a *Seeker*, then additional progress on the *Pathway* slows or stalls completely.

PREVENTATIVE FUNDAMENTALS

There are three key *"hot buttons"* that are most troublesome to a *Seeker* (*processing* their way along the *Pathway*) if they are ignored rather than handled (or *confronted*). The primary handling of these areas is covered in former lessons of this *Professional Course*; but by illuminating them here together, we can discuss their application as "preventative fundamentals" for everyday life and *upper-level processing*. They are:

1. A *break* or *upset* in the *"Flow-Factors"*—enforced or inhibited *communication*, *likingness* and/or *agreement*. [see *Lesson-7*, "Eliminating Barriers"]

2. A *"Human Problem"*—present-time *attention* (*presence*) is occupied fixedly elsewhere (and outside one's own control). [see *Lesson-4*, "Handling Humanity"]

3. A *"Hold-Out"*—*attention* restimulated the area, usually because someone else *almost found out* about it. [see *Lesson-6*, "Escaping Spirit-Traps"]

In *Traditional Processing*, a *Professional Pilot* would usually check these *"hot buttons"* at the beginning of a *Formal Session*. This rudimentary step has obviously required additional training (to employ effectively in *processing*) and is not included in the basic *Formal Session Script* given as an example at the very start of the course. Some *Seekers* even make a practice of checking for these *"buttons"* in a daily *session*.

You might check these over now, at this point of the lesson, and see if there is anything present that requires handling. Usually, a *Solo-Pilot* does not have to check these over at every *session* if they are *processing* often and moving along the *Pathway* well. If something does come up in life, you will usually be able to recognize it and handle it—but, of course, in *Traditional Processing (Co-Piloted)*, a *Professional* would check these over each time just to ensure the best results possible from the remainder of the *session*.

As a general rule: if a *Seeker's attention* is presently "stuck" or "fixed" on one of these *"buttons"*—an *upset, problem* or *hold-out*—then additional *"subjective" processing* techniques are not going to be effective unless they are targeting that area directly. On the other hand, *"objective"* techniques (involving *selectively directing attention* on *objects/environment*) can often aid in "cooling down" enough to allow for *subjective processing*.

For example: you can apply the *preventative fundamentals* to help "clean up" something that just happened in regards to one of the areas targeted by the former lessons of this *Professional Course*. Each lesson addresses specific areas and usually includes a general technique that may be adapted for "single-use" handling of a specific thing that has just been "stirred up" or manifested as a "barrier" of some kind.

Usually, when an area has been completely *run* to a stable *"release-point,"* a *Seeker* is mostly free from *fragmentation* in that area—so, when something related does come up, they are able to simply *confront* it (and apply knowledge of that area) to the situation to *prevent* any persistent issues. Of course, if a *Seeker* finds their *attention* fixed on some troublesome area, then a review of these *fundamentals* may be in order.

Now, eventually, as a *Seeker* begins to handle more and more areas, some of the earlier *"release-points"* will *destabilize*. This is normal—and it occurs because the range of *Awareness* (or *availability* of *"knowables"*) is expanding. More of the data from "hidden areas" becomes accessible—and, of course, this means there is much more available to *process* in an area then when having experienced the previous *"release-point."*

For example: during a *Seeker's* first "pass" through the *Professional Course* material, the *processing* on a certain area is *run* to a *"relief"* or *"release-point"* involving what is currently available—and if that technique would be *run* further, it would be *overrun*. However, as a *Seeker* works further and deeper into various areas, when you go through an additional "pass" through the material, there is more available now to *run*—and those earlier *processes* become useful again for progressing further into the various areas.

To advance our example further: supposing we take the area of *"protest"* or *"problems"* for this example; a *Seeker* might first *run* through the material as a basis for stable *Beta-Defragmentation*, using the *viewpoint* of *this* single *"Human"* lifetime for *processing*— and upon completion, will experience a great deal of increased *Awareness* and *Self-Actualization*. But as more of the *Backtrack* comes into view, concerning a long span of existence as an *Alpha-Spirit*, there are much "higher" *viewpoints* that may also be handled further. Having already maintained the first stable point, these more "advanced" areas are more accessibly reached.

In *Traditional Piloting*, if a *Seeker* slows or stalls on the *Pathway*, it is up to the *Pilot* to determine why. Of course, really it is only the *Seeker*, themselves, that knows—even if they don't think that it is the reason why. But, it is still a *Pilot's* responsibility to discover it for all concerned, and often this involves probing various areas directly (often with the assistance of an electronic *biofeedback device*). In *Solo-Piloting*, a *Seeker* is left to discover underlying *fundamental breaks* or *upsets* themselves—*identifying* and *confronting* them directly.

In addition to targeting the three key *preventative fundamentals*, a *Professional Pilot* will usually "probe" the following list: Has anything been *protested*; Have you committed a *Harmful-Act*; Is anything being *suppressed/repressed*; Has there been a *false accusation*; Has anything been *invalidated*; Has someone *enforced* an *evaluation*; Are you *holding-out* a *communication*; and, Has there been a *misunderstanding*. Somewhere within all of these areas, the underlying reason for slows and stalls will usually be found.

There is one additional aspect that we have not yet covered in this course, mainly because it pertains specifically to *systematic processing* itself, and not one of the typical areas of *fragmentation* (such as with previous lessons). As mentioned briefly earlier: our communication of *Systemology* is not "fool-proof"—and individuals do make mistakes in *processing*. Therefore, *"processing errors"* are the additional reason why a *Seeker* might run into trouble now and then when using our methodology. This makes for an almost unique form of very light *"potential fragmentation"* that is specific only to our own *Seekers*.

One of the most frustrating *processing errors* is "looking for something that isn't there." There may be a previous mistake or misunderstanding that leads to it; but for whatever reason, a *Seeker* may decide that something is "there" or "true" when it isn't. For example: deciding that there is a *break* in a *Flow-Factor* when there really isn't, so *running* that for "relief" turns out to make them feel more *confused* or *upset* than before—because obviously that wasn't *it*, &tc. This generates greater unnecessary *turbulence*.

Processing Errors are not the "end of the world." If you can simply *"spot"* the point where you went wrong, or got a *"wrong indication"* from a *process* or exercise, then you can "back up" and try again. This can happen even to the most expert among us. All available data or reasoning will *indicate* that one thing or another *is it*, but then once that is done or handled, the original issue seems to remain. This occurs all the time in our everyday world, especially when attempting to "repair" a complex *"machine"* or *system*.

CONFRONTING THE PAST

Individuals tend to *reactively withdraw* from painful and undesir-able experiences—and any *associated data* and *"mental imagery"* attached to them. Failing to face—or *confront*—the contents of an *incident*, an *Alpha-Spirit* stores *fragmented* and *irrational* data re-garding the *facets* and *factors* therein.

Defragmenting an *incident* properly using *systematic processing* re-quires all of the experience, knowledge, and *Actualized Awareness* earned from the *first nine lessons* of this *Professional Course*. This ability is, however, the main staple or *end-goal* of *Systemology Level-4*. It also potentially opens up new areas of the *"Backtrack"* (the part of one's *Spiritual Timeline* that is *"past"*) for *processing past-lives* (as that data becomes more available).

Confronting-the-Past is so critical that it represents an entire tier or *processing level*. Hidden within those past "negative experiences" are *considerations* and high-power *Alpha-Thought "postulates"* about what the *Alpha-Spirit* is no longer *willing* to *Be, do,* or *have.*

This *unwillingness* leads to *creating* automatic/reactive *"mental mechanisms"* (*spiritual machines compulsively created* with *personal energy*) to handle these undesirable *facets* and *factors* on a *continu-ous* and *"unknowing"* basis. This is quite detrimental to a *god-like Spiritual Being.*

"Unconfronted" (*"repressed"* and *"Not-Known"*) areas become *frag-mented* "blind-spots" for the individual. This is, for example, why an individual tends to find themselves in certain circumstances again and again. By not *confronting* it properly—or seeing it *"As-It-Is"*—the effective actions to appropriately handle the situation are not taken. *Reactive withdrawal* (based on earlier *imprinting*) *fragments* one's *clear perception* when it seems like the undesirable thing might happen again—or even when an *associated facet* of it is present.

In *Systemology*, we often use the word *"restimulated"* to describe the *present-time* "activation" of past *fragmentation*. By this, we mean when the circumstances of ordinary life "stir up" or "bring to the surface" the *fragmentation* of past incidents—and, of course, this occurs *outside* or *apart* from the *Self-determinism* of the individual; it happens automatically or reactively. The individual tends to bring the whole past *incident* into the *present*, and often feels (or gets a sense of) the effects of that *incident* as if it has just happened again.

While some of our *vocabulary* and *theory* is specific to *Systemology*, this phenomenon we are describing, has been known to humanity for thousands of years. [Described in the *Systemology Core* volume: *"The Tablets of Destiny Revelation."*] It is observable in practice; not simply a theoretical idea. Our words, like *imprints*, and explanation of a *Backtrack*, *are*, however theoretical constructs—but they are workable. Using our ideas provides a greater understanding and more effective results for an individual *Seeker* than the *"figuring-thoughts"* collected over thousands of years on various types of *spiritual mysticism* and *psychotherapies*.

Our *Systemology* also requires a greater level of *Actualized Awareness* (and personal *realizations*) in order to "go all the way" with it. It requires *Knowing* things with *certainty* that are not so easily accepted when simply "taught" or "instructed." This is why we encourage *Seekers* to actually "apply the work to their life" and not simply treat this as a purely "intellectual philosophy" or "faith-based religion."

For example: the basic underlying *truth* is that the *Alpha-Spirit*—the individual, themselves—*creates* their own *Universe*, their own *mental* and *emotional states*, their own *automated-mechanisms*, and beyond this, even carries around an entire library of painful and undesirable *mental images* that requires suspending *attention* (and *creative ability*) on the *past* in order to hold onto them. Of course, most of this now operates *unknowingly*—however, our complete *Beta-Defragmentation* procedural regimen (as outlined in the *Pro-*

fessional Course) is designed to bring various areas *knowingly* back under the *control* of the *Seeker*.

It is that "library of old pictures" that we are most concerned with when "*confronting-the-past.*" Essentially, the effects of these *mental images* only linger because they are of things that have not been *confronted*. This is the only reason for an *Alpha-Spirit* to keep them around *compulsively* and *continuously*, rather than just *Knowing* them and being able to "*create*" and "*destroy*" the *mental imagery* at will.

The *Alpha-Spirit* carries *pictures* of *incidents* that have not been properly *confronted*. Since the individual did not fully "face" the *pain* (or *loss*) of what happened, there is now a lack of data (and *Awareness*) in that area. Yet, at the same time, the individual doesn't want it to happen again. This puts them in a position where they are afraid to actually "forget" what happened, but at the same time, still doesn't want to have to deal with it *knowingly*. This is how *fragmentation* can form: *continuously creating* a *reactive mental image* that is allowed to operate without *knowingly* "looking at" or *controlling* it.

Our earliest *systematic* approaches to this phenomenon were (and are) effective, but incomplete. Those *Seekers* using the information in the first edition of "*The Tablets of Destiny*" in exclusion (without the "*Crystal Clear*" companion volume) tended to focus too much attention on vigorously "*erasing*" the "*pencil lines*" of *fragmentation* (metaphorically speaking) to the point of nearly tearing through the paper.

The more recently developed "*upper-grade*" approach emphasizes increasing one's *ability* to simply *confront* the past incidents to a point where they are no longer inhibiting or affecting present/future *considerations* and *action/creation*; the "*pencil mark*" is no longer "vivid" or "bold" enough to affect whatever else we might do with the page. It is still kind of there, the *incident did really* happen, but we don't care.

True *"Beta-Defragmentation"* occurs when a *Seeker* works with this material long enough to where they no longer need—and therefore can "toss out" or "cease creating"—whatever *mental machinery* they have set up as *Alpha-Spirits* to handle undesirable data. A *"Self-Honest"* individual is no longer *unknowingly compulsively creating* a "library of old painful pictures" that are *reacting* outside one's own *control* and *Self-Determinism*. Of course, this itself is not the *"ultimate"* state of *Ascension* that we refer to—but it is a *necessary prerequisite state* for *Ascension*.

DEFRAGMENTING THE PAST

Whether we call it *Beta-Defragmentation* or *Self-Honesty*, this state we are describing is the first stable point of *"metahumanism"* available to a *Seeker* on the *Pathway*. By this, we mean that it *exceeds* the limited range or parameters that we otherwise use to define the "Human Condition." An individual who is free from operating *reactively* based on *past imprinting* is no longer a *standard-issue* "Human."

Our *systematic* approach works on a *gradual-gradient* to *reach* this state. A *Seeker* simply begins with lighter— easier to *confront—incidents* before *processing* the more difficult ones. Doing this allows a *Seeker* to *gradually* increase their *Actualized Awareness* and *ability-to-confront* the nature of *Existence "As-It-Is"* until greater, older, and deeper areas of one's *past* come into view for inspection (*analytical recall*).

There is also no reason to become overly concerned with searching out *"The (Ultimate) Incident"*—some single instance or *incident* that might be used to explain *all* of a *Seeker's* underlying *fragmentation*. While it is true that there are some *incidents* on the *Backtrack* that have been more detrimental to one's spiritual decline than others, that type of pursuit is not nearly as effective for a *Seeker* (in terms of stable results) as what we present for the *Professional Course*.

Although there are *systematic* guidelines that we follow, there is not only one single way of which to *run* a *process* for *Confronting-the-Past*. In fact, this is one of those practices that we have found evidence of in many times and cultures throughout history. The difference is our *systematic processing methodology*. And for our present purposes, this course will emphasize a technique that is appropriate for *Solo-Piloting*. While *running past incidents* is sometimes handled earlier in practice with *Traditional Piloting*, *Seekers* must have raised their own *Awareness* with *all* former *processing levels* (from previous lessons) before *Flying-Solo* in this area.

The data for these first *processing steps* should be recorded in your *notebook* (or *"flight log"*).

A. *"Identify the incident."*

Simply *"spot"* the *incident* on your *track* (*past*) and identify it with a label. To begin with something simple and presumably *non-restimulative*: we'll use the example of *"getting the mail yesterday"* for our lesson.

B. *"Spot when the incident occurred."*

If you can't *"spot"* the *timing* directly, determine it as closely as you can. Consider ranges of time and see what feels correct. If it happened a long time ago, sometimes it is easier to consider it in terms of its order-of-magnitude—such as how many "years" or "decades" it's been. Be as concise as possible with these *steps*, because it will contribute to how clearly you can bring an *incident* into view *"As-It-Is."* For our example: you might *spot* the *"exact time that you started walking to the mailbox."*

C. *"Spot the duration of the incident."*

Determine "how long" (time-frame) the *incident* lasted. This helps improve *recall* accuracy. For our example: we might *spot* a *"ten second duration of walking to the mailbox."*

D. *"Spot the location of the incident."*

Determine exactly "where" the *incident* took place. At the very least (for older or more obscure *incidents*), try to get a sense of the "direction" and "distance" it was from your present location. In our basic example: we might have *spotted* the "*location of the mail-box.*"

E. "*Spot the size-of-space that the incident took place in.*"

This helps provide a background for which the *incident* will be viewed. Did it take place in a large space outdoors or a small room or did it happen across many miles, *&tc*. For our example: we *spot* the "*spatial area of the yard between the front door and the mailbox; a walked path of approximately 100 feet.*"

F. "*Close your eyes.*"

This obviously completes the preliminary "written" *steps* of the *process*.

G. "*Move to the beginning of the incident.*"

Direct your *attention* to the *actual* time when the *incident* happened. *Recreate* the "beginning" point of your experience of the *incident* as if the scenery is all around you—as if you were actually there again. In our example: one might *spot* "*standing at the front door looking out across the yard at the mailbox.*"

H. "*Move through the incident.*"

Do your best to completely *re-experience* the *incident*. See what you can *notice* or *perceive* about everything that happened in it.

I. "*Open your eyes; write down what happened, what you noticed or perceived.*"

Record everything that happened and anything you perceived. This is done to see it as "external" and "separate" from *Self*. For our example: we might record that we "*walked across the yard, reached the mailbox, opened it, and found it empty.*" Other perceptions (things noticed) might include the "weather," "smells," if any "animals" were present, *&tc*.

J. [*Repeat Steps F through I.*]

Run through the *incident* again. When you get to the "written" part, simply record any *new* details that may have been noticed this time.

After *running* the *incident* a couple times, a *Seeker* should be able to tell if the *fragmented charge* associated with the *incident* is being "*processed out,*" or if it is getting more "*solid*" and difficult to *confront*.

If the *fragmented charge* is being properly "*processed out,*" then some *new* details *should* emerge that you didn't notice on the first *run*—or some of the details might "rearrange" themselves to be more in line with *truth*. In either case, the actual "impact" of the *incident* should begin to weaken; it now seems unimportant.

If, however, the *incident* seems like it is getting more "*solid,*" "*heavier*" or more difficult to *run*—as in the act of *confronting* the contents is making you feel worse about it—then there is *earlier* data (such as from a *prior incident*, or even an earlier "beginning point" of the same *incident*) that needs to be *run*.

Our "*mailbox*" example is not very *restimulative* of *fragmentation*; but if it were, and if we were having a hard time of it, then we would look for either an earlier "beginning point" of that *incident*, or an earlier *incident* that also included a "*mailbox*" or "*going to get the mail.*" Perhaps the real "beginning" of the *incident* was the "*decision to go out to get the mail, while still inside*" and not "*standing outside the front door.*" This would change the *spatial area, duration* and other data required to properly *run the incident*.

As you *confront* an *incident* from the past using *systematic processing*, it should become easier. If that is not happening, and you cannot find an earlier beginning-point of the *incident*, then simply "looking at" or "stirring up" this *incident* has actually *restimulated* earlier and deeper *fragmentation*. The only solution is to *spot* or *identify* the earlier one. This happens quite often *unknowingly* in our everyday encounters.

Since an *earlier incident* is being *restimulated*, its virtual presence is usually right there almost "behind" the *incident* you are looking at. If it doesn't come into view right away, usually you can get some sense, feeling, or impression, about it. It is even possibly something more obscure, such as from an *earlier lifetime*. When just starting on *Systemology Level-4*, the only reason to *run incidents* from *past-lives* is if it comes up while *processing your present life*. Whenever you find an earlier "beginning-point" or *incident*, start with "*Step-A*."

"STEP-K" : DEFRAGMENTATION

It is important to note that *heavily "charged" fragmentation* has a tendency to "*alter*" and "*change*" the apparent "*IS*-factor" of an *incident*. We call it "*fragmentation*" for a reason; and it has a tendency to *distort* the data that is *recalled* or *considered* when viewing (or attempting to *confront*) a strongly *charged* or *fragmented* "*imprint*" (or "*mental image*" of a past experience). [This is why we covered the nature of *Existence* for the first half of this lesson.]

If the *content* or *sequence-of-events* keeps shifting around, you are still dealing with "*It-As-Changed*." This means you should keep *running* it until the data seems to settle one way or another. Although the general rule is to *run* whatever comes up in *processing*, the data being perceived may be quite inaccurate while the area is still "*heavily charged*." If an individual is still encountering an *incident* "*As-Changed*," it cannot be properly *confronted* "*As-It-Is*."

You are finished with this *process* when the *incident* (or *earlier incident*) is completely "*processed-out*." This means there is no longer a "*mental charge*" or "*emotional turbulence*" attached to the *incident* (or *thoughts* and *considerations* of the *incident*). The event, itself, is not actually "forgotten" or "erased from memory." The *mental imagery* can still be *recreated* and *dissolved* at will—but the key here is "at will." There is no longer a reason for the individual to continue *unknowingly compulsively creating* it as *reactive-fragmentation*.

For example: if our personal survival were dependent on the information found in books, than an individual would only compulsively carry around those physical copies of books that they did not already *Know*. If the material were memorized (or *Known*), there would be no need to refer to the book—and no reason to insist on keeping that book close. If, however, the contents of one of them were of such a nature or magnitude that the person could not fully *confront* or truly *Know* it, then they would want to have it as a future reference. This may be a bit on the fantastical side as an example, but it is relevant for our purposes.

The increased *Awareness* and new *realizations* are the *systemological* version of *Knowingness* from this example above. As more areas can be brought up to the level of *Known*, more freedom and clarity is to be had by the individual. The *Alpha-Spirit* is able to regain more *Awareness* and *control* over its own *automated-mechanisms* and the *fragmentation* they have generated. More *Awareness* is available because less "units" of *spiritual attention-energy* are being suspended in the *compulsive continuous creation* of *fragmentation*.

As a *Seeker* progresses further on the *Pathway*, it is possible to get nearly instantaneous *defragmentation* simply by actually *looking/confronting* with high-power *Actualized Awareness*. The *Seeker* feels the "relief" just on proper inspection of the *fragmentation*. But more often than not, a *Seeker* will have to repeatedly *run* an *incident* completely through and *spot* whatever they can *notice* about it several times before true *defragmentation* occurs.

Once a *Seeker* reaches an *end-point* on the *processing* as given, the final (more advanced) step of the *defragmentation* procedure is to *"spot any considerations or decisions that were made at the time of the incident."* This means to actually *identify* the *Alpha-Thought/postulates* that you have carried along ever since—and which are likely to have *unknowingly* influenced the manner in which you experienced *reality* thereafter. This may be too far of a reach on your first pass through this course, but it is an important part of its mastery.

If *Solo-Piloting*, it is better to work with *"uncharged" incidents* until you are well-practiced in this technique. Then you can work up to actual *"pain incidents,"* but starting with the more trivial ones, like *"stubbing your toe."* In such an example as the *"mailbox"* or *"stubbing a toe,"* there is likely to be many, many occurrences of a *similar incident*. You don't actually have to work backwards one-by-one with each time. You try to *spot* the *earliest incident* available for *recall*; and then in *running* that, you may discover a new *"earliest"* one. This can happen several times as you gradually increase your *ability-to-confront* for that entire area.

The random aches and pains of life usually stem from, or are more intense because of, *restimulation* of an *earlier incident*. Rather than handling or *processing* a "condition," this technique may be applied to *incidents*. A *Seeker* doesn't use *systematic processing* to *run* "an aching toe," but instead, "impacts to the toe." *Running* a simple pain might even bring you to *confront* a major "accident" or "event," in which case, you handle it as an *incident* with the full procedure given in this lesson.

The "heaviest" *imprint fragmentation* actually developed within *this present lifetime*, will be during periods of *"biological unconsciousness."* This area is very closely tied to *accidents*, *physical trauma*, and *medical operations*. It is also tied to very deep far-running *chains-of-fragmentation* that are likely to extend far beyond *this* life. Improving your *Awareness* and *ability-to-confront* any area will bring more into view; just be patient with it, don't push too hard, and this technique might just surprise you.

ON HANDLING "LOSS"

In the beginning, an *Alpha-Spirit* is quite *god-like* in its *creative ability*. Eventually, the lack of *"value,"* with things being so easily *created* and *destroyed*, becomes rather boring. When something *can* be *created* into "infinity," there is nothing to make "this" more valuable or interesting than "that" until we begin to assign "*significances*" to things. So, "*uniqueness*" enters in as a *consideration*.

"Loss" emerges much earlier, further back, on our personal *track* than the idea of "pain" (as a source of potential *fragmentation*). "Pain" is really a *reactive-mechanism* that developed afterward to give warning of a "potential loss." In the case of a "*Human Body,*" the *Alpha-Spirit* has forgotten its own *creative ability* and is dependent on the "body" (failing to be able to *recreate* one at will) and thus there are many *reactive-mechanisms* in place to "protect" against its "loss." But, keep in mind, the *Alpha-Spirit* is still *unknowingly creating* these *mechanisms* too.

On additional, more advanced, "passes" through this course material, you can see just how far the technique can go. For example: you can *run "past-deaths"* from the *viewpoint* of *Self as an Alpha-Spirit*, and *confront* the *loss* of "bodies." This starts to open up wider *considerations* (and *Awareness*) of the *Backtrack* prior to *this* lifetime. You can also *process-out* the *loss* of former civilizations, and even earlier *Universes*, that you once experienced.

You may discover many other potential applications for the *systematic processing* of this *Professional Course* that will assist you in "*reaching further.*"

Your only limit is Infinity.

LESSON TEN
"LIFTING THE VEILS"

SPIRITUAL BEINGNESS

In addition to gaining greater *control* over the *"Mind-Body"* systems inherent in the experience of the *"Human Condition,"* a *Seeker* will also have increased their *Awareness* on the *truth* of *Being* an *Alpha-Spirit* that is *having* a *Human* experience—and we now continue further in this area for *Systemology Level-4*.

Previously in these *Professional Course* lessons, we began suggesting the idea that an individual, in their native state, is an *Alpha-Spirit*—a unit of *Spiritual Awareness* that is not, itself, actually located in *this Physical Universe (Beta-Existence)*. Of course, the *Alpha-Spirit* can *think* or *consider* that it is located in *space-time*; and it can also *think* that it *must only* operate from the specific *viewpoint* of the place it *thinks* it's located.

An *Alpha-Spirit* is without "form" or "mass." It is *Pure Awareness.* In order to experience *"pain"* (and therefore a potential *"loss"* of a *Body*) an *Alpha-Spirit* must *"postulate"* or *"strongly consider"* its own *Beingness* as one-and-the-same with a *Body*; by this, we mean to fully *"identify"* *Self* with a *Body*.

At basic, we cannot actually be "harmed," but once we *consider* that we *must have* a *Body*, then the *fragmentation* about *"losing"* a *Body* follows shortly thereafter. Once an *Alpha-Spirit creates* (or *projects*) *energy-matter* (by *intention* and *attention*) to operate in a fixed position (*identifying Self* and establishing a sense of *ownership* there), the *Being* (as a *Spiritual Awareness*) can now be "located" in that position—thus it can be "harmed" (or "hit") in that position. Of course, eventually that position deteriorates (or is "smashed"/"collapses") and they must establish a new position to operate from.

As this pattern of activity continues, much of an individual's own history (*Spiritual Timeline* or *Backtrack*) consists of "abandoning unsafe" or collapsed locations and frequently "leaving stuff" behind. This systematically weakens the *spiritual power* of the *Being*, because the native ability to *have* and *create* "*space*" is governed by the ability and willingness to *reach*—in this case, to literally "*knowingly reach*" rather than "*reactively withdraw*" from *locations*. In this regard, we are really speaking about two distinct *levels* of *existence*—*Beta* and *Alpha*.

There is, of course, the *Physical Universe* (*Beta-Existence*). An *Alpha-Spirit withdraws* from those "unpleasant locations" and "painful positions" until they can no longer operate outside (or apart from) a *Body*. Then, there is the individual's own *Personal Universe* (or *Alpha* state), which includes what some refer to as the "*mental universe*" equivalent of *space-time energy-matter*.

The term "*mental*" is not systematically appropriate for our purposes. Often, the term is meant to simply mean "non-physical" or "metaphysical," but we prefer to define the "*mental universe*" as part of an *Alpha* or *Spiritual Existence*—one that is composed of the *Alpha-Spirit's* own *creations*, in contrast to the *shared creations* of a *Shared Universe* (or *Beta-Existence*).

"LOCATIONAL" PROCESSING

On a higher *Alpha* level ("higher" than what we perceive with the *Human Condition*), the *Alpha-Spirit* is operating from a fixed location in its own *Personal Universe* along with the "*controls*" and "*systems*" for experiencing *Beta-Existence*. In the "video game" of *Life*, the "controls" and "system hardware" are not elements experienced directly as part of the *game* that is appearing displayed on the *screen*. These elements are *exterior* (and in many ways "superior") to the more apparent *Game Universe* that is being experienced from an *interior viewpoint*.

We focus primarily on *Beta-Defragmentation* in our course instruction, because the truth is that many of the cycles, tendencies, behaviors and patterns that are observably exhibited in *this* lifetime and *this Universe* are indicative of the same type of things we find further and earlier on the *Backtrack*. The same *fragmented mechanisms* involved (or attached) with *withdrawing* from *unsafe locations* in *Beta-Existence,* also applies to what is taking place (at an *Alpha* level) regarding an individual's own *Personal Universe.* As the *Arcane Tablets* from both *Mesopotamia* and *Hermetic Traditions* relay: *"As Above; So Below."*

In *running* the following *processes,* simply *consider* whatever "comes up" or "feels right." The *locations* indicated in these *processing command-lines* ("PCL") do not have to be restricted to *this Existence,* though a *Seeker* is likely to start with answers that are relatively closer or more familiar. As an individual stops *compulsively withdrawing* from *occupied locations,* more *spiritual ability* comes back under their *control.* This *systematic processing* can be taken much further on additional passes through the *Professional Course*—but initially it may be applied lightly when first introduced to this *processing level.* For each of the following PCL, *spot* many *locations*—and allow for the one's that may not make sense (or seem rational) to come up as well.

A. *"Spot some places where you would be safe."*

B. *"Spot some places where a parent (or guardian) would be safe."*

C. *"Spot some places where a child would be safe."*

D. *"Spot some places where a lover (or companion) would be safe."*

E. *"Spot some places where a teacher (or leader) would be safe."*

F. *"Spot some places where lifeforms would be safe."*

G. *"Spot some places where valuable belongings would be safe."*

H. *"Spot some places where energy would be safe."*

I. *"Spot some places where ideas would be safe."*

J. *"Spot some places where a spirit would be safe."*

K. *"Spot some places where it would be safe to keep a god."*

The following PCL-formula concerns *communicating* an *intention* or *control* to a particular *"part of the body"* as a general *process*. There are other applications of this too—such as to assist *attention* in managing "pain" or "difficulties" with a particular *"part of the body."* A *Seeker* will likely start with "physical locations" as their answers, but as an advanced practice, they should try to work up to *considering* the locations within their *Personal Universe.*

This PCL is a formula that is used for multiple *processes* by inserting different "terminals" into the blank space. First start with *"bodies,"* using the terminals: "HAND"; "FOOT"; "STOMACH"; "GENITALS"; "HEAD"; "EYES"; "EARS"; and "BRAIN." Then use terminals for *"roles/identities,"* such as: "LOVED ONE"; "AUTHOR-ITY FIGURE"; "ANGRY/EMOTIONAL PERSON"; "DANGEROUS CREATURE"; "VICTIM"; and "GOD."

The PCL-formula is:

"From where could you communicate to (a) ---?"

Now, regardless of how many PCL you used (or how many answers you came up with) for the *locational processes* given previously (in this section), *run* the following two in *alternation* a few times.

A. *"Spot some places you are willing to be."*

B. *"Spot some places you are willing to not be."*

Our final set of *processes* for this section involves another PCL-formula. This time we will apply it to *"concepts"* rather than *"terminals"* (with mass). The PCL uses the word *"create"* to mean everything from lightly imagined *visualizations*, to vivid *mental images*, to really manifesting something in *Beta-Existence.*

The PCL-formula is:

"From where could you create ---?"

First start with some basic concepts: "A PICTURE"; "AN EMO-TION"; "A MENTAL MACHINE"; "AN AUTOMATED REACTION";

"FRAGMENTED CHARGE"; "A PAINFUL FEELING"; and "A MIND." Then, apply some of these advanced concepts (or wait until a second pass through this course): "BEING TRICKED"; "TRICKING ANOTHER"; "BEING ABUSED"; "ABUSING ANOTHER"; "BEING BETRAYED"; "BETRAYING ANOTHER"; "BEING TRAPPED"; "TRAPPING ANOTHER"; "A GAME"; "AN IDENTITY"; "AN OBJECT"; and "A REALITY."

CREATION AND KNOWINGNESS

At the top of the *Alpha* "scale" of *Existence*, there is only "*Creation.*" Whereas the goal of *Beta-Existence* (or a *Games Universe*) may be "*Survival,*" in our true native *god-like* state, our *Alpha* purpose is solely "*To Create.*" Even at a *Human* level, the act of "*creating*" is incredibly satisfying and desirable. But this is not the only *state* or *condition* that is available to (or able to be experienced by) a *god-like Alpha-Spirit.*

The *Alpha* state of "*Creation*" includes a level of *Knowingness* that extends far beyond simply "*knowing about*" something. It is an "all-pervasive" (nearly absolute) type of *Knowingness* concerning our own *creations.* But, as *god-like Eternal Beings*, this state does not provide us with much in terms of "*games.*" When we *know-all* and *see-all*, there is no "randomness" or "chance" possible for us to experience—there is no "*game.*"

As *Spiritual Beings* with a near-*infinite* existence, maintaining a position at the very top of the *Alpha* scale became "boring." We wanted to also be able to "*go into*" our *Creations* and "*play*"—to be able to actually experience our *Creations.* As such, we developed a specialized type of "*Not-Knowing*" that could be practiced from our *god-like* state in order to make our experience of existence— our experience of our *Creations*—"more interesting." This works out just fine until we find ourselves with a bunch of *rigidly fixed Not-Knows* that are now outside of our *control.*

Therefore, just below *"Creation"* (at the top of our *Alpha* scale) we have a specialized state of *"Not-Know."* This, at least initially, is *Self-Determined* (as a *"postulate"* or *Alpha-Thought*) to intentionally "block out" the perfect *Knowingness* (that one begins with as *Creator*) in order to make experience of *"games"* possible—in which we now have something to *"Know-About"* or *"find out about."* The scale further continues gradually downward—but in this lesson, we are not concerned with "scraping the bottom."

The following *processing exercises* help a *Seeker* get more perspective on this area by *knowingly* practicing it—even if only conceptually. In *running* these: you can *"Not-Know"* something that you really *do* know, something you *don't* know, or even something you are *unsure* of. For example: with *"objects,"* you might already know how much it weighs, but in practice, decide you don't; or perhaps you might really not know anything about when it was made, and provide that as an answer. You can *"Not-Know"* (*Step-B*) many things about a particular "terminal" within the same *process*. These practices are intended more for educational purposes than to reach a specific *end-point*.

NOT-KNOWING: OBJECTIVE

A. *"Look around the room and spot an object."*

B. *"Decide to 'Not-Know' something about it."*

NOT-KNOWING: POPULATED AREAS

A. *"Look around and spot a person."*

B. *"Decide to 'Not-Know' something about them."*

NOT-KNOWING: ZU-VISION

A. *"Close your eyes; Spot individual things (that might be) in the room."*

B. *"Decide to 'Not-Know' how they look."*

The previous *process* (above) is an advanced practice that *may* help improve *"ZU-Vision"* perceptions (*exterior to the Body*). This *process* is effective because, from within the *Human Condition*, you are already *"Not-Knowing"* how things appear in *Beta-Existence* unless the *Body's Eyes* provide you this sensory information. This is, of course, not true for the actual abilities of the *Alpha-Spirit*, but it is really how things are (or seem) when operating from within fixed *viewpoints "interior to"* the *Human Condition* (or *Body*).

NOT-KNOWING: CREATION

A. *"Create a mental image of an object."*

B. *"Decide to 'Not-Know' who created it."*

And finally, whenever *running "Not-Know,"* even if only in conceptual practice, it is best to *end-session* with actual *recall processing* (of something *Known*). Therefore, you can complete this area by: selecting something that you can easily remember (such as your favorite food); then alternately decide to *"Know"* and *"Not-Know"* it a few times. Then select something else and repeat this.

METASPIRITUAL SYSTEMOLOGY

Experience of the physics and material knowledge of *Beta-Existence* is only one part of what we may apply *Systemology* to. At this point on the *Pathway*, it should be obvious that an individual is much more than just a *Body*—and even the *Mind-System* is "metaphysical," far beyond the scope and reach of "neural-brain sciences." Some of this lesson is based on material that first appeared in a volume of our *Systemology Core*, titled: *"Imaginomicon."*

The *spiritual side* of *Systemology* offers a *Seeker* more "substance" than what is found in any *religious tradition* or style of *mysticism*. As many *Seekers* (with previous backgrounds with these other methods) have found, our *Systemology* is also far more effective for increasing and/or rehabilitating our true *Spiritual Awareness*. Of course, this does not occur all at once; and it often requires handling our experience of *Beta-Existence* first.

An *Alpha-Spirit* is not simply operating a *Beta-Body* in this *Physical Universe* from its native *Alpha* state. To say so is an over-simplification. The experience of *this* version of *Beta-Existence* has such "apparent solidity" to it because it is the "accumulation," "compactification," and/or "condensation" of all the former versions of *Beta-Existence* (*Universes*) that have since "collapsed" into *this* one —a sort of "common denominator."

The *Alpha-Spirit* also does not "directly" experience the *Human Condition.* The "*communication*" of that experience—and its *control* by an *Alpha-Spirit*—is filtered and relayed through many levels of *metaphysical or "mental" machinery, energetic fields and beams*, and various layers of "*subtle*" or "*astral*" *bodies*, all of which remains *continuously created* and *actively running* in the present (from use in previous existences), though the individual has become so entrapped by the *Human Condition* that they are no longer aware of this.

Therefore, what we would *consider* an "appropriate" *Body* for the *Alpha-Spirit* in this *Beta-Existence* is also a "condensation" or "greater solidification" of former similar *considerations*. What is maintained from "previous bodies" (existences) "*collapses-in*" on the *Human Body*. We use the expression "*entrapped by*" to be less *restimulative* (of *fragmentation*), but it is more systematically correct to say that the "layers" and "levels" of *Awareness* (and *energetic-stuff*) maintained by an *Alpha-Spirit* (as its own perceived *identity*) progressively "*collapses in*" on the *Human Body/Condition* for various reasons (for example, when trying to "protect the body" from danger/loss).

Ideally, to *control* a *Body*, an *Alpha-Spirit* would remain entirely *exterior* to it—simply extending a *reach inside* using *intention* (and/or *energy-beams*) to operate one. In its native (*unfragmented*) state, the *Alpha-Spirit* maintains most of its "*energetic-machinery*" completely outside of a *Beta-Existence* in order to keep it from getting too mixed up with *energy-fields* and *reactive-systems* specific to the *Body* (or "*genetic-vehicle*" operated in *Beta-Existence*).

But, again: various circumstances cause *Awareness* to *"collapse in"* or *"snap-in"* on the *Body*—and the *Alpha-Spirit* begins to associate *Self* with a *Body* more and more until their *viewpoint* tends to be *"inside it,"* because the individual believes they *are* it. The *"snap-in"* effects tend to be *energetically turbulent,* which causes a type of *fragmentation* that inhibits an individual from *"ZU-Vision"* (or *"spiritual sight,"* where an *Alpha-Spirit* is not dependent on sensory perception from within the *Human Condition*).

The *Alpha-Spirit* began *"outside"* (or *"exterior to"*) *Beta-Existence*— and any associated *genetic-vehicles*—prior to being *"snapped-in"* to it. This is the first area that requires *defragmenting* in order to lift some of the veils that distort our *clear view* as *Alpha-Spirits.* For this, we directly apply to this area the *systematic processing techniques* a *Seeker* has already developed proficiency with throughout this *Professional Course.*

To begin, we will provide a list of *keywords* (or *"buttons"*) and a basic *processing-structure* (or *formula*). We are targeting any incidents that relate to *"going interior"* in some way, in order to *relieve* some of the *fragmentation* attached to *automatic/reactive "snap-in"* effects. Select one of the *keywords* from the list that seems the most interesting—and once that is *run* well, choose another to *process* until you *"feel good"* in regards to this area.

The *keywords* are: "GO IN"; "PUT IN"; "WANT TO GO IN"; "MUST GET IN"; "CAN'T GET IN"; "KICKED OUT"; "BE TRAPPED"; "FORCED IN"; "PULLED IN"; and "PUSHED IN."

In *Traditional Piloting,* the language structure of a PCL is easily modified *in-session* to apply to various lists such as these. A *Seeker* will notice that the following *processing-structure* is worded to apply to, for example, the first, second, and third *keyword* on the list, but would not make sense for some of the others. For *Solo-Piloting* a *Seeker* should write out the proper PCL for a specific *process before running it,* if it does not fit the *structure-formula* as printed in the lesson. [For example: *"have the feeling or sense that you must get in."*]

1. *"Recall making someone ---."*
2. *"Recall being made to ---."*
3. *"Recall someone making another ---."*
0. *"Recall a time when you decided to ---."*

If any of these *"buttons"* seem particularly *"charged"* or *"turbulent"* (and do not seem to be *defragmenting* with *Analytical Recall*), use the techniques and training from the previous lesson (*"Confronting the Past"*) to *run* the *"chain-of-incidents"* or *"chain-of-fragmentation"* that is holding up (or suspending) *Awareness* in that area. The *"chain"* is typically specific to *one* of the four *"circuit-types"* treated in the *process* (above). [If necessary, a *Seeker* may also run a PCL using *"imagine"* in place of *"recall."*]

METAUNIVERSAL SYSTEMOLOGY

In addition to the *"snap-in"* effects generally associated with an *Alpha-Spirit's* own *Awareness* regarding a *Body*, there is a similar area of *fragmentation* attached to *"collapse-of-space."* By this, we mean literally the phenomenon of *"energy fields"* and *"space"* collapsing-in-on the *Alpha-Spirit*.

At the upper-most levels of our *Systemological Philosophy*, the *"collapse-of-space"* phenomenon is most strongly tied to experiencing the *"Collapse-of-a-Universe."* As one might guess, this is a very *turbulent* area of *spiritual fragmentation*. The magnitude of potential *"loss"* or *"sadness"* connected to this area far exceeds anything an individual is likely to newly experience in *this* lifetime. But as with other areas of *fragmentation*, there are everyday situations where this might even become *restimulated*.

This area represents some of the earliest *fragmentation* we have ever experienced—and it is innate to each and every one of us. The first time it was strongly *imprinted* is during the original *collapse-of* the *Alpha-Spirit's* personal and individual *"Home Universe."*

Since then, there have, of course, been other experiences that later reinforced the *fragmentation* to a point where the individual no longer is *willing* to be personally *responsible* for "*Creation-of-Space*."

This present lesson is, in many ways, a direct continuance of the "*Locational Processing*" and "*Creation-of-Space*" exercises introduced in *Lesson-5* ("*Free Your Spirit*")—and elsewhere, in our "*Imaginomicon*." These previous exercises depicted the basic structure of *space* (or a *Universe*) as a "*cube*"—defined by *eight* "*corner-points*" that "*anchor*" the boundaries (or dimensions) of that *space*. An *Alpha-Spirit* maintains "*anchor-points*" that they use to define their own personal perception of *space/creation*.

Just as there are experiences of one's own *point-of-view* ("POV") "*snapping-in*" on a new "*plane*" of *space*, or a new "*mass*" (or "*Body*"), so too are there *imprints* of when *space* and *energy-matter* —and the very "*corners*" or "*anchor-points*" of a *Universe* itself —"*collapse-in*" on the individual.

This area is *defragmented* similarly to the *systematic processing* from the previous section (above). Essentially, a *Seeker* is looking to *run* this area as *incidents*, using their training and experience with procedures in this and previous lessons. We are targeting different *incidents* than before. Rather than a *viewpoint* ("POV") being *snapped-in* on some *space* or *mass*, the reverse is *processed*.

The "*hot buttons*" (or *keywords* or *phrases*) used for *defragmenting* this area include: *incidents* of *space collapsing-in* on the POV; the world *closing-in* or *folding-up*; *energy imploding* or *collapsing-in* on the POV; *corner-points collapsing-in* or *snapping-in* on the POV; sudden "*uncreation*" or *folding-up* of all *forms/space*; the environment/*space caving-in* on the POV; being *pulled back* from; *falling away* from; and the *sense* that everything is suddenly becoming *unreal*.

Handling this area properly may be too advanced of a step for some *Seekers* on their first pass through materials of the *Professional Course*; get what you can from it and spend more time in this area when you come back to it. As an additional note from the *upper-levels* of *realization*: it is always the *Alpha-Spirit* themselves that *collapses* their own *space*. Although environmental stimuli and pressure (or trickery) from others may get you to do it, you are the only one that can actually *snap-in* your own *"anchor-points."*

SPIRITUAL ENERGY & BEAMS

In the beginning, an *Alpha-Spirit* operated on pure *Alpha-Thought* and *"postulates"* — *creating* things and making things *just happen* by *intention* alone. But below this level, as we began to *Share Universes* with others, we made *reality-agreements* concerning the necessity of using *energy* to *create* and *change* things.

In terms of the *Human Condition*, or the operation of a *"genetic-vehicle,"* the *Alpha-Spirit* tends to direct *energy-beams* into the head, into the back of the neck, into various spots up and down the spine, and the *"solar plexus"* or stomach region. There is also the tendency to wrap a long coil-like beam around the mid-section (or torso), which can then be used to abruptly *pull* the body from perceived danger.

When operating a *Body* using *energy-beams* while *"exterior to"* it, there isn't much "kick-back" from the *Body* since everything is handled remotely. If, however, one's own POV is fixedly *"interior to"* the *Body*, then even the presence or impact of these *energy-beams* alone can produce intermittent or sporadic headache sensations and stomach discomfort, *&tc.* — an entire variety of *"pings"* on our *Awareness* become possible.

Using experience from previous *training levels*, some *Seekers* have learned to provide *"relief"* for these specific types of *"pings"* by determining what the *energy-flow* of the *beam* is, and reversing it; or by repeatedly alternating the *intention* to *increase* and then *reduce* (or *reverse*) it, until it becomes under greater *control*.

Although handling *energy-beams* directly is a more *advanced level* of practice, it is never too early to start *considering* what is *actually* happening with one's own personal *energy* and *Awareness* when various *incidents* occur—or *have occurred*, when *running incident processing*. We have been handling this area *lightly* and somewhat indirectly all throughout the *Professional Course*. This makes *recognizing* it much easier as a *Seeker* progresses further along the *Pathway*.

Since the *Alpha-Spirit* maintains a continuous *Spiritual Timeline* (or *Backtrack*) that extends through many previous *Universes*, it is a reasonable conclusion that we are certainly not limited to a *location* within *this* version of *Beta-Existence*. Of course, without heavy *fragmentation*, we are well more *aware* of this fact—and we *know* that we are *reaching into* our *Creations* and *Games Universes* in order to operate and experience things.

Given that the *Alpha-Spirit's* own actual (*Alpha*) existence is *exterior to* any experienced *Beta-Existence*, it does not have any reason to not operate from an *interior viewpoint* to have a better "*game*" experience. One might even note the sheer increase in "*first-person perspective*" video game popularity in the past few decades.

Although a *god-like* being might, at first, be quite able to operate multiple "*bodies*" or "*games*" simultaneously, eventually more and more *Awareness* "*collapses-in*" on *one* specific *lifeform* or *genetic-vehicle*. By this point, the native *spiritual ability* to simply "*recreate*" a *Body* at will has also diminished. The *Alpha-Spirit* is now able to feel "*loss*" when a *Body* is destroyed—hence the "new-found" *intentions* to *protect* one.

Then, finally, when the *Alpha-Spirit* "*identifies*" or "*associates*" *Self* too closely with the *Body*, experience of *pain* is possible—mainly to give warning about a *potential loss* to (or of) the *Body*. When beings realized how unpleasant it is to occupy the "center" of a *Body*, they created "*control centers*" to operate from that are close to, but stand apart from, the *Body*.

This is only partly what we mean in *Systemology* regarding a "*Master Control Center*" (of the *Alpha-Spirit*) versus a "*Reactive Control Center*" (that is specific to a *Body*). And while the types of mechanisms are "*metaphysical*" in nature, they *are "energetic-constructs*" that have *form* or *mass*—and therefore have "*control points*" or "*anchor-points*" that define its nature and the *metaphysical space* it occupies. These "*points*" also have a tendency to "*snap-in*" or "*collapse-in*" on the *Body* during certain *incidents*—and with the type of *effects* or *fragmentation* we have been describing throughout this lesson.

When *running* some of the earliest *incidents* in this area, the *Alpha-Spirit* will still be operating from a point *interior-to* a *Body*, while simultaneously projecting "*operating/control-points*" that are *exterior-to* the *Body* (but placed closely around it for protection). If a *Seeker* doesn't have much "*reality*" yet on this idea, it may be taken up on future passes through the *Professional Course* material. However, the *systematic-sequence* that may be *processed* is (*run*) as follows:

1. *Operating from an "interior" control-point.*
2. *Operating from an "exterior" control-point.*
3. *The "exterior" control-point "collapses-in" to the body.*

As an additional advanced practice for this area: *consider* (or *imagine*) some ways that an *Alpha-Spirit* that does not have a "*Body*" could get "into trouble." See if you can *spot* a "*chain-of-incidents*" to *run* that involves something you came up with for this. So, *Imagine* some ways, pick something that *could* be a part of a particular *chain*, and *run* it. Don't worry about *heavy implants* (or "*conditioning*") for this *processing level*. If they are run into now, just "*recognize*" that such things *do* exist.

"ZU-VISION" & PERCEPTION

Increased *Awareness* and *perception* is an inherent part of progressing on the *Pathway*. However, research has suggested that the *fragmentation* associated with *"going interior"* or *"snapping-in"* (as described previously) should be *"processed-out"* prior to placing a direct emphasis on *"ZU-Vision"* or (*"spiritual perception"*). The *fragmentation effects* from the first area prohibit and inhibit development or rehabilitation of the second.

Before completing *Systemology Level-4*, a *Seeker* is presented with a series of *systematic processing* exercises and techniques that directly encourage increasing *perceptions* of *"ZU-Vision"* (or *"Spirit Vision"*)—which is to say, placing emphasis on using *viewpoints* that are *exterior-to* the *Body* (*Human Condition*).

As with the presentation of earlier similar exercises in this course, it is expected that a *Seeker* will initially experience a mixture of *real perception* and *imagination* when first practicing in this area. It is important to not *invalidate* your progress by comparing what is *perceived* in practice with what you might *see* using eyes of the *Body*. The development (or rehabilitation) of *true spiritual ability* occurs gradually.

As a good general "warm-up" for this type of work, repeatedly *alternate* the following PCL with *eyes closed*, until you have *some* degree of certainty regarding your *perception* *"exterior-to"* the *Body*. Some *Seekers* prefer to practice these types of *processes* while lying down. [*Eyes closed; blindfolded if necessary.*]

A. *"Spot three points in the body."*

B. *"Spot three points in the room."*

This next *process* is best run outdoors (with *eyes open*). Repeat it many times until you can really get a sense of *"space."* Once you have some *reality* on the exercise this way, perform the same steps with *eyes closed*.

A. *"Spot two objects."*

B. *"Notice the distance between them."*

C. *"Get a real sense of the 'space' between them."*

There is a very ancient and much misunderstood spiritual practice referred to as *"Journeying to Other Planets"* or *"Traversing the Star-Gates."* In many ways, it is meant to extend a *Seeker's attention* further and further away from the rigid entrapment of the *Human* experience on *Earth*. It is meant to provide a *realization* of a simple truth—although much additional *fragmented* esoteric hype has been attached to it.

In *Lesson-5*, we introduced part of this technique using *imagined secondary viewpoints*. This time, we will be using the actual *planets* in this solar system—just as the ancient practice suggests. We will make a more literal exercise out of it, using the *physical planets* themselves, rather than *consider* anything esoteric about what they might symbolically represent.

This *process* is *run* repeatedly in series several times for each *"planet"* (*terminal*) listed: "EARTH"; "THE MOON"; "MERCURY"; "VENUS"; "MARS"; "JUPITER"; and "SATURN." Once you have practiced what is here to the point of an increased certainty on your *spiritual perception* (*ZU-Vision*), you can later apply additional steps from *"Locational-POV Processing"* in *Lesson-5* to advance this exercise further. [*Eyes closed; blindfolded if necessary.*]

A1. *"Be above ---; looking down at it."*

A2. *"Spot three points on the surface."*

B1. *"Be inside ---; occupying its center."*

B2. *"Spot three points interior to it."*

The next exercise is best practiced outdoors. It requires simply walking around; but as you do, focus on getting a sense that you are remaining in one position and moving the *Universe* beneath and around you. Then, add the earlier steps (from further above) of *"spotting objects"* and *"noticing the distance/space"* between you and them. After this is accomplished with ease, start to *alternate* between the idea of you *"moving through the Universe"* and *"moving the Universe around"* you (keeping with one idea for a few minutes before switching to the other).

To complete this lesson and *processing level*, we include several exercises that first appeared in the original *"Wizard Training Regimen"* outlined for the *Systemology Society* many years ago.

• With eyes closed: *Imagine* a *duplicate* (identical *facsimile-copy*) of your presently owned *"Human Body"* out in front of you. Make a copy next to it. And another. And several more. When you have about eight or so, push them together into a "ball" (or single *mass*) and *collapse it* into nothing. *Imagine* another *duplicate.* Make a copy next to it; and many more copies. Then push them together into a ball and toss it away. Continue this step until you feel more comfortable easily *creating (imagining mental imagery)* "bodies."

• With eyes closed: *Imagine* a *duplicate* of your present *body*, seeing it as "ideal" and "healthy." Then unmake it. Make it again; then unmake it. Repeat this several times.

• While looking into a mirror: *Get the sense* that there is "something there"; then *get the sense* that there is "nothing there." Alternate between these *considerations* repeatedly several times.

• With eyes closed: *Imagine* a busy or crowded place (mall, depot, or street corner). Decide to *be there* (place your *point-of-view* in a specific location). Look around and *spot* "terminals" (*objects* and *people*) and "*motion*" in the scenery. Practice this for multiple locations (preferably until there is an increase in actual perception).

• Use the location you like best from the previous exercise: *Imagine* making a *facsimile-copy* of your present *"Human Body"* to use as a *point-of-view* ("POV" or *viewpoint*). Then, unmake the *Body*, but remain in the location as an *Awareness*, continuing to look with your *point-of-view* there. Repeatedly alternate making and unmaking the *Body*, but still being able to "look out at" the surrounding scenery.

• Perform the previous exercise: *Imagine* a *facsimile-copy* of the *Body* out in front of you, but this time using a *point-of-view* "outside" the *Body* to look around at the location. Then use a *point-of-view* from "inside" the *Body* to look around at the location. Practice repeatedly alternating between these two *viewpoints*.

• Perform the previous exercise (in full), but this time adding: *Get the sense* of other persons acknowledging your presence when they are near (or walking by), even if they don't look at the *Body*.

• Select a basic solid "object/shape" (*pyramid, cone, cube, sphere, &tc.*) for practicing the four previous *locational* exercises as a cycle: *Imagine* using the *object* as your body; making and unmaking, alternating *viewpoints*, *spotting* other *terminals* and *motions*, and receiving acknowledgments ("*hellos*" *&tc.*) for your presence.

• Perform the previous exercise, but this time adding: unmaking the *Body/Shape* and *point-of-view* from one spot and making it again at other spots in the location. Also: *Get the sense* of moving the *Body/Shape* like a "playing piece" of a *game* (across a *game board*).

• Perform the previous *locational* exercises as a cycle; do this completely (as a complete *process*) for each of the following "*Body*" *terminals*: a duplicate of your *present body*; an *elderly body*; a *child's body*; a *different gendered-body* than your present one; a *sparkly-cloud* of *silvery-white energy* with small *golden balls for eyes* (as a *body*); and finally, using your *Awareness* as a *point-of-view* (with nothing added as a *body*).

The final exercise (below) requires the ability to *conceive of* or *maintain a sense of* "centering" or "focusing" your own *Awareness* as a *viewpoint* (or POV) a few feet *behind* the *head* of your present body. In addition to this, a *Seeker* should be able to *expand their perception of space* (as an *Awareness*) to include the *Body* from this "distance." This is a progress checkpoint on the *Pathway*. If this is above a *Seeker's* present *reality* and *ability*, than this is a suitable point to cycle back to the beginning of the *Professional Course* for an additional "pass."

Often, this *sense* of "operating from behind the head" is best practiced in a "public" place at first (because of the amount of *terminals* and *motion*), but it may also be practiced when performing your everyday routine (when it is safe and appropriate to do so). As a *systematic process*, it is best to allocate a specific period of time for its actual "practice," although there are some *Seekers* that eventually find that they start to *knowingly* operate in the mode more as they progress further along the *Pathway*.

In this practice: as you "walk the body" around a public place, you will still use its eyes for general spatial-orientation, but while *having a sense* that you are larger than the *Body* and encompassing it from a few feet behind the head.

To further this practice: as you *observe* or *look* from a single stationary location; use the *body's eyes* to *spot* some *terminals*, then *spot* some *motions*. As you do this, *get the sense* that the *body's eyes* are sensing the *images/scenery* and sending this information to the *brain* and *control-center* of the *Body*—which, in turn, is *relaying* this "*communication*" along some kind of channel to *you* as the *Spiritual Awareness* behind the *Body*.

As an advanced step: attempt to *ignore the perceptions* received and relayed by the *Body*, whether eyes opened or closed, and *get a sense of* or *imagine* observing the *Body*—and surrounding *scenery* (*terminals* and *motions*)—using *only* your *viewpoint* as an *Awareness* behind the *Body*. Then alternating between this step and the previous *stationary location* step.

This completes *Systemology Level-4*.

LESSON ELEVEN
"SPIRITUAL IMPLANTS"

BEYOND BETA-DEFRAGMENTATION

Depending on the extent to which a *Seeker* has *run* the former *processing-levels*—and, of course, the amount of *highly-charged past incidents* that have been *confronted* and *defragmented*—the first primary stable state of *"Beta-Defragmentation"* is often experienced by the end of our former lesson for *Systemology Level-4* (in combination with all previous lessons).

Systemology Level-5 begins with this present *Professional Course* lesson; and this *processing-level* serves as an additional *"booster"* or *"stabilizer"* for basic *"Beta-Defragmentation"* –and– as a "primer" for *"upper-level"* work (as preparation for additional *Advanced Ability* training and *Alpha-Defragmentation*).

The benefit from this *level* only comes after accomplishing the previous work on the *Pathway*. The material in *this* and *future* lessons does not replace the training, skill and increased *Awareness* that is gained *only* by properly following the *systematic procedure* outlined for *Systemology Level-0 to 4*.

Once sufficient training and skill in previous *processing-levels* is attained, a *Seeker* may even continue to use any of those techniques (as they best apply) to handling whatever comes up in *sessions* (and *Life*) thereafter. Increased *certainty* (and *willingness*) in being able to "handle" (or *confront*) *"Existence"* is what opens up more "data" from the *"Backtrack"* (or one's personal *"Spiritual Timeline"*) for *processing*. [Much of the material given in this lesson is based on lecture transcripts (from *March 2023*) published as *"Systemology: Backtrack"* or *"Advanced Systemology: Academy Lectures, Volume 6."*]

Too often it has been found that there is a certain "glamour" or "excitement" attached to exploring the *Backtrack*—or the "uncovering" of *"past-life"* memory, *&tc.* But, often times, an individual does not realize the true reason that this information is hidden; they don't realize just how much *attention* (or in essence, *energy* or *effort*) has been placed on "blocking it out" (as a *Spiritual Being* with *that* ability).

Our methods promote getting a "handle on" what has taken—and *is* taking—place in *this* "*Lifetime*" before chasing down accumulated *fragmented imprints* from countless "*past-lives.*" However, at a "higher level" of *realization*, we *recognize* that an individual or *Alpha-Spirit* has *one continuous* "*Spiritual Timeline*" that is merely being *experienced* as relatively separate *forms* or *bodies* ("*incarnations*") and in various *Universes*.

In addition to *processing* exercises and techniques, the *upper-levels* of the *Professional Course* also begin to emphasize a more "expert" handling of *systematic processing* (*sessions*) in general. This not only allows a *Solo-Pilot* to handle more "advanced" material on their own, but also increases the *certainty* that a *Seeker* might have to *Pilot* another.

As the title of this course suggests, a *Seeker* completing it will have earned the minimum requirements to *professionally* or *vocationally* provide (*Pilot*) "*Beta-Defragmentation*" procedures (*Systemology Level-0 to 4*) for others—to assist the *Self-Actualization* and *Ascent* of other *Seekers*. While this is not the primary goal of this *Professional Course*, most *Seekers* find that "wanting to share it" is a natural side-effect of its completion.

"*Helping Others*" is actually one of the few activities—apart from *systematic processing*—that effectively takes the "weight" off the type of areas we *defragment* on the *Pathway to Ascension*. But, of course, before we can most effectively and clearly *help* others, we must first have *helped* ourselves reach a point of such personal stability that we can safely and appropriately extend our *reach downward* to lend a *helping hand*.

IDENTIFYING "SOURCES"

To handle the *Backtrack*, advanced concepts of *"Spiritual Implanting*," or even to have a better understanding of *Life-experiences* in general, it is important to understand *who* is doing the *looking*. The more *"metaphysical"* techniques and exercises of previous *processing-levels* are intended to provide a *Seeker* with greater *certainty* on *"Self as the Alpha-Spirit"* (rather than *identifying Self* as a *"personality-package"* or any kind of *"body"*).

By this point along the *Pathway*, it is critical for a *Seeker* to focus directly on improving their *"Awareness as a Spiritual Being"* in order to "break free" from more "attachments" associated with artificial *"personalities"* and *"identifications."* This is what provides stability to a *Beta-Defragmented* state, which in turn allows for "higher-level" advancement in those "areas" that are now more available due to previous *processing*.

Anything that an *Alpha-Spirit* carries (*compulsively continuously*) along its own *"Spiritual Timeline"* is a product or *creation* of its own *considerations* and *Alpha-Thought* (*"postulates"*)—however much these might also have been coerced, *enforced*, "implanted," or otherwise influenced, from "outside" (*"other-determined"*) *sources* along the way. A "high-power" *Awareness* that *this is* what's happening is *"Step-One"* to its *undoing*.

To reinforce this idea, we direct the first exercise of this *processing-level* toward *spotting sources*. By *"source,"* we mean what is *"originating"* or *"causing"* something—for example, a "place" or "point" of origin, from which something comes (or is *communicated*). At this time, we are mainly interested in simply *running* through the *considerations* that occur in *processing* much more than whether or not we are "correct." This may be practiced with any "condition" or "situation" that interests the *Seeker* (or else, an area that is being *defragmented*).

A. *"Spot something that could be a 'source' for ---."*

B. *"Spot something that is probably not a 'source' for ---."*

By this point, a *Seeker* should be familiar with the use of *"spotting"* in a *processing command-line* ("PCL")—but, this time, we are not restricting our *"range of view"* to only that which is physically present in the room or environment. We mean to specifically direct *attention* and *Awareness* onto something—including, for example: *"mental imagery"* or *"memory"* that is not, itself, a part of *Beta-Existence* (*reality* that others also can *view*), but are *actual* experiences (or even reflections of the *Backtrack*) that the individual (*Alpha-Spirit*) "sees" (or *Knows*).

An inability to *identify* even a single *source* indicates a high degree of *fragmentation* (or low-*Awareness*). By this, we mean too, that they are unable to be "in communication" with their environment. Such individuals appear (to others) quite "out-of-touch" since they are obviously also misappropriating *cause* for *Reality* and *Existence*—and may also be still quite enamored by other *"mystical"* and *"magickal"* traditions or beliefs.

A *Seeker* that is unable to *identify anything* as a *"source"* or *"cause"* of *anything* is likewise unable *to be* a *Self-Determined "source"* or *"cause"* themselves (at least *knowingly*). Of course, people are *causing* things all the time; but their level of *Awareness* determines just how *Self-Directed* those actions (or behaviors) truly are.

We began with a more direct *process* "about something" for practice—but a more *general process* for this area is:

A1. *"Spot a source."*

A2. *"Notice something about it."*

B1. *"Spot a non-source."*

B2. *"Notice something about it."*

There are some *Seekers* that quickly spot *"Self"* as the only *"source"* (as an *end-realization*) and then move off from the *process*. Not only does this miss the point of the *process*, it is also inaccurate. A *Seeker* that *only* identifies themselves as a *source*, will experience wide fluctuations between states of *"mania"* and *"guilt."* Of course, when high-power *Awareness* is applied to *perceiving sources* and *causes*, the effects of *fragmentation* easily fall away.

For additional practice, perform the following exercise in a public place. These *"PCL"* may be used fully on various different *"terminals"* in a single *process*, or you may alternate the (*"B"-Steps*) on a single *terminal*. Just continue *running* it (as it works best for you) until there is a *realization*, things seem *"brighter,"* and/or you are *"feeling good"* about the *process*.

A. *"Spot a person (or lifeform)."*

B1. *"Notice something they are causing."*

B2. *"Notice something they are not causing."*

And finally, whether *processing* indoors or outdoors, select a small object or portion of wall to look at. Then *run* the following *considerations* alternately *"on it"* until there is a change in *Awareness*, or one of the *end-points* just described previously (above). Most of the upper-level *processes* are simply *run* until such points. Additionally, in this case, as you cycle through these PCL multiple times (and eventually using different objects, *&tc.*), focus on increasing the sense you have of *"Step-1"* each time you *run* the *process*.

1. *"Get the concept that you are creating it."*

2. *"Get the concept that others are creating it."*

3. *"Get the concept that no one is currently creating it."*

IDENTIFYING "WHAT-IS"

Even when only nearing the completion of an effectively *run Beta-Defragmentation* regimen, a *Seeker* should arrive at the *realization* that the *Alpha-Spirit*—the I-AM, the actual *Self*—is *"non-local"*; that its true *Beingness* is not *actually located* within the *"space-time"* of *this*, or any, *"Beta-Existence."*

While experiencing the *Human Condition*, an *Alpha-Spirit* has rigidly fixed its own *"point-of-view"* (POV) to a *singular viewpoint* that *is* *"local."* But, any *identification* related with *"this viewpoint"* as being the same as "one's own personal *Beingness"* is purely a matter of *associative considerations*; which makes it *"real"* (for purposes of experiencing a *reality*), but not *"actual."*

In *actual fact*, the *Alpha-Spirit* has never really left from its original *"static"* position as a *"unit"* of *Spiritual Awareness* (that is *aware* of being *aware*). It has, however, manifested many *"postulates"* of *creation*, many *"layers"* of *consideration*, and of course, *reality-agreements* regarding our ability to be in *"communication"* with others in a *"Shared Universe."* All of this contributes to what we call *"Existence."*

The pattern of accumulating *fragmentation* and subsequent *condensation* or *solidification* of *Universes* (*Existence*)—as reflected on the *Backtrack*—suggests that we are experiencing cycles of a "downward spiral." Of course, this is not the "bedtime story" that we reveal to *Seekers* early on the *Pathway*. It is not meant to be discouraging—but it does indicate how critical *defragmentation* is to the *"Spiritual Being."* Because it will go on *existing eternally*, but only *experiences* what it considers *to exist*.

Just as a high-power *attention* on *actual "sources"* and *"causes"* will elevate a *Seeker's Actualized Awareness*, so too do we find an additional boost in stability and certainty when a *Seeker* becomes *more aware* of the *existence* of the *actual* in contrast to the *non-existence* of the *non-actual*. As with *"sources,"* this is important for high-level *defragmentation* of *"imprints"* and *"implants"* (stemming from further on the *Backtrack*).

In *Systemology*, when we talk about *reality*, and *control* over our experience of *existence*, we are actually talking about *"creation"* and *"what-Is"*—*"Is"* with a capital *"I"*—such as our handling of the *"IS-factors"* (discussed previously in *Systemology Level*-4). Very often, a *Seeker's* personal conception of *"what-Is"* will be greatly affected by *where* on the *Backtrack* (the *imprint* or *incident*) that much of their own *attention* "units" are *still* "stuck," "hung up" or "suspended" on (as *fragmentation*). The nature (or qualities) of such an *incident* (if not the *incident* itself) might even *resurface* by alternating the following PCL:

A. *"What is?"*

B. *"What isn't?"*

In *Traditional Piloting*, a *professional* uses such a *process* to monitor and increase the "present-time" *presence* that a *Seeker* is currently applying to higher-level *processes*. Is there *presence* really in the *past?*—having more *attention* stuck on past incidents; past *IS-ness* (such as with *"psychosis"*). Are they actually *present* in the *present* —or just continuously *"stuck"* in the *present* (as with *"neurotic"* tendencies)? The ultimate goal would be to get a *Seeker* to collect all their *attention-units* up and have them back under *Self-directed control.*

A *Seeker* should answer from a *present-time viewpoint* for several cycles of the *process* before moving off of it. What is more likely to happen, is that a *Seeker* will start with *present-time existences* and then move further *back* in the *past* before coming back to the *present-time existences.* It is also possible that a *Seeker* may start with distantly *past existences* first. In any case, the *process* ends when *only present existences* are coming up (or being consistently repeated).

For an additional (less "esoteric") *process*: *consider* the following, in alternation.

A. *"What must be a part of your existence?"*

B. *"What must not be a part of your existence?"*

If you were to add another *"circuit"* to this *process*, you would also run *"must/must not"* for *"another's"* existence. And finally, we apply an *objective process*, using a large object or wall—spending a few minutes holding the *concept* suggested by each of the following PCL as they are cycled.

A. *"Get the idea that it is there."*

B. *"Get the idea that it is not there."*

C. *"Maintain both ideas simultaneously."*

Having done all this, let's step out a bit further by applying this practice to another *objective process*—which also uses the walls of a room. Once seated comfortably, with eyes open, *look at the wall*, then *visualize* that it is "transparent" or "see-through" (which is to say *create* an "empty space" where the wall currently is). Now, *imagine looking through the wall* and seeing what is on the other side.

You can practice this with different walls in the same room—then different rooms—*spotting* things through the wall. Don't be overly concerned with accuracy (or checking to see if you are right); just work with this as best you can until you "feel good" about your practice.

"CONDITIONS" & "SIGNIFICANCES"

When we speak of a specific manner or state of *Beingness*, *IS-factors*—and various other things that either are essential for, or modify, *existence*—we are referring to *"conditions."* For example: *fragmentation*, itself, is a *condition*. It restricts the nature of things in one system and causes or allows things in another. Another example: the experience of *Beta-Existence* (this *Universe*) is composed of many *conditions*. It requires a specialized package of *fragmentation* in order to compulsively and fixedly experience the *Human Condition*.

Properly *identifying conditions* and how they were handled (meaning, what we did about it, and/or what *considerations* we still maintain from it) is critical for *defragmenting* areas that are not targeted directly in former *processing-levels*. The reason for holding off on this until now, is that high-power *Awareness* is required.

Although it may not occur, it is very likely that a deep source of *turbulence* will *resurface*. If and when this happens, do not change *processes* just because an undesirable experience is encountered. The *fragmentation* or *imprinting* must be *confronted* and *processed-out* by repeatedly using this same *systematic process* (if it is this *process* that *resurfaced* it). If necessary, you can always "soften the impact" of *confronting* by noticing (or *spotting*) something about what you are handling, and alternately *spotting* something in the room (*present*).

Many *conditions* and *considerations* may have to be *run* in this *process* until the "target" *imprint* or *incident* is *identified*. It is possible that it does not *resurface* during the first *session* that this *process* is applied. The ultimate goal here is to *identify* the *single imprint* or *incident* (for an area) that the *Seeker* is still *persistently creating* or *carrying* into "present-time." The concern for this type of *processing* is not how long it is *run* while still "missing the target," but whether it gets *overrun* beyond the purposes described in this section.

The *general process* is:

A. *"Spot an existing condition."*
B. *"What have you done about that?"*

In *Traditional Piloting*, the first PCL is only repeated if a *Seeker* needs to refocus *attention* for answering the second. By this, we mean an individual is likely to have *done* many things about a *single condition*. When all of the *considerations* for a *condition* are *processed*, you can then *run* it on another *condition* within the same *session*. If *identifying* and *spotting* (in the *process*) are not sufficient in directing high-power *Awareness* toward *defragmentation*, then a more direct PCL may be added to the second part, such as: *"What part of that can you confront?"*

Although a *Seeker* is able to remember (or *recall*) many different past *incidents*—and while there would seem to be many *imprints* that require *defragmentation*—there is likely to be *one* truly significant *imprinting incident* that has been persistently carried forth into the present, upon which all other *imprinting* is "stacked up" on. This originating *incident* would then act as a *"platform"* for other *fragmentation* to accumulate on. Such *"platforms"* are more accurately referred to as *"implants"* in our *Systemology*. The subject of *implants* will be taken up in more detail later on.

As a transition point into our next area of focus, *run* the following similarly to the previous process (above):

A. *"Spot a deeply held belief (you have)."*

B. *"What have you done about that?"*

When we handle *imprints* and *implants* in *processing*, we are not "erasing" or "forgetting" things—but we are taking the "weight" off the *significance* and *importance* that has been *compulsively* or *reactively* assigned to things, which is to say, how *much* we are *choosing* to allow ourselves to be the *"effect"* of our own past.

One of the ways to practice with this lightly is to reduce the *"force"* that a *Seeker* associates with *action words* such as *"hit,"* *"smash,"* *"break,"* *"explode,"* *&tc.* As a *process*, you would say the word while visualizing a *mental image* of the *force* (such as the action taking place); then say the word and decide that nothing is associated with it—until the word no longer produces an *automatic reaction* (or *mental image*) of *force*. Note that this does not eliminate the "meaning" of the word.

From experience with *systematic processing*, another way we know of freeing up *considerations* is to knowingly exaggerate "both sides" until the handling of it is under greater *control*. Concerning *"significances,"* we might alternate the following PCL on a particular *incident*, *event* or even an *object*:

A. *"Decide that it is important."*

B. *"Decide that it is not important."*

It should be understood that our intention here is not to eradicate all *significance* or "meaning" to the things in a *Seeker's* life, or that they enjoy, *&tc*. What we want to do is rehabilitate the full power of personal choice over the *consideration* of *importance*—the kind of "fluid freedom" a *"god-like being"* would ultimately have.

SPIRITUAL IMPLANTING

A full handling of all *spiritual implanting* on the *Backtrack* exceeds what is treated in *Systemology Level-5*. These *implants* typically do not inhibit reaching a basic state of *Beta-Defragmentation*. However, they are sometimes encountered while *"Confronting-the-Past."* Therefore a *Seeker* is introduced to the subject here, so that they are better prepared for when they *do* show up in *processing*— whether now, or when handling more *advanced processing* directly for future progress on the *Pathway*.

In our previous publications—such as the *Systemology Core* volumes—*"implants"* are defined as "platforms" or "patterns" on which *imprint fragmentation* accumulates. This is, however, only part of the story, so to speak. As such, this whole subject may be above the level of *reality* for some *Seeker's* to *confront* on their first pass through the *Professional Course*. It is still a critical area for advancement.

A *Spiritual Being* typically experiences an *"implanting incident"* upon *entry into* a specific *Shared-Games Universe* (*e.g. Beta-Existence*). Other *implanting incidents* show up on the *Backtrack*, particularly in the *"between-lives"* areas (that are intentionally) "hidden" from typical view (*Knowingness*). Such *implants* originate from (or are *created* by) more *"advanced civilizations"* than what is currently present on Earth. *"Implanted Universes"* likely originate from a previous *Universe* (preceding versions of a *Beta-Existence*).

During the long span of our *spiritual existence*, we have maintained various positions of *control*. *Implants* are often efforts to simply *"enslave"* beings. *Implants* have also been used (by *advan-*

ced civilizations) to *"condition"* (or *"enforce"*) a certain *"ethics"* (or else to *"stop"* an individual from *doing* various things). There is evidence on the *Backtrack* to indicate that we have *all* been on *both sides* of this activity at one time or another.

In regards to this present *Universe, implanting* is commonly conducted by "hitting" a *Being* with *"electronic waves"* in order to give an "impact effect" to the *commands* or *command-lines*. These are *not* the same type of *"command-lines"* or "PCL" used in basic *Beta-Defragmentation* procedures (*Levels 0 to 4*); but we do *knowingly run "implanted commands"* (and *confront* them with high-power *Awareness*) for our advanced *processing-levels*, in order to take some of the influential "impact" or "force" *off* of them.

In order to differentiate between types of *command-lines*: an *"implant-command"* is systematically referred to as an *"implant-item"* or *"command-item"* for *processing* (because it is really neither a *"terminal"* nor a *"concept"*—but we *process* it the same, because it does have *energetic-mass* attached to it).

Most *implanting incidents* are quite distantly "old" on the *Backtrack,* and their remaining *"command force"* (by themselves) is actually quite weak. It is primarily the additional *imprinting* on that *platform* (or *foundation*) that makes their *impact* or *fragmentation* seem more "vivid" when *restimulated* in daily activities—or if encountered unexpectedly when *incident-running* at earlier *processing-levels.*

Generally speaking, only *"advanced" Seekers* will go directly "looking for" *implants* to *run* (as part of *Alpha-Defragmentation*). Material regarding *"implants"* is classified *"advanced"* (by the *Systemology Society*) because the contents of *implant-incidents* and their *"command-items"* should not be "casually scanned" without actually *systematically processing* (*spotting* and *confronting*) what was being *implanted*. Otherwise, we again find unnecessary *restimulation* on the *Pathway* (which can inhibit progress).

Implants and *implant-incidents* can strongly affect clear *"past-life re-*

call." By its very nature, *all fragmentation* "distorts" our view—but we differentiate between *"implants"* and *"imprints"* for good systematic reasons. For one: *implants* are essentially *"artificial imprints"* that originated intentionally from an external/outside (*"other-determined"*) *source*; they are not an inherent part of everyday life. Everyday life is how *imprints* get layered on top of (or superimposed over) a preexisting *implant*. This is why/how *"imprint-chains"* are manifested and carried from one experience (or *incident*) to the next; and one *"lifetime"* (or even *Universe*) to the next.

Another distinguishing characteristic of more significant (or *"heavy"*) *implanting-incidents*, is the amount of *confusion* and *surrealism* experienced—which far exceeds what an individual would find in "ordinary" circumstances (even at relatively "higher" levels of *existence*). Geometric *shapes*, complex *patterns*, false *imprints* (*events/incidents*), and misleading *time/dates*, are all common components (or *facets*) used to give the *"command-items"* a longer lasting impact; because they would otherwise not persist as long as they do.

When a *Self-Actualized* individual maintains high-power *Awareness*, the residual effects of most *basic implanting* can be reduced to the level or degree of an annoying "commercial" or advertisement (relatively speaking). This is what prevents additional *fragmentation* from "stacking up" *again*. But, of course, in lower-*Awareness* states, these *"implanted suggestions"* (from an outside *source*) more greatly affect our *"postulates"* and *considerations* about *"what-IS"* (as treated earlier in this lesson prior to this section on *implants*).

DEFRAGMENTING "IMPLANTS"

There is a basic technique for *"defragmenting implants"* available to *Seekers* that have successfully reached this point of the *Professional Course* (and development on the *Pathway*). We will give more explanation of these "steps" as we go along—and there are other ways to advance this work further (using a *biofeedback device*)—but for present purposes, the basic method of handling *implants* (in *Level-5*) is:

A. *"Spot (or Imagine) an implant-item (in the listed sequence)."*
B. *"Confront it until it ceases to have an effect."*

And based on what we have learned (and used) throughout this course, the easiest way for a *Seeker* to accomplish this as a *systematic process* is to alternately:

A. *"Spot the command-item."*
B. *"Spot something in the room."*

This basic technique is really not an exercise in *"desensitizing"* or reducing impact of "words." It is important to note that *"command-items"* are/were *not implanted* in "English" (or any language of *Human* speech). Therefore, the best we can do for *processing* is to "approximate" their meaning (and *get a sense* for their original intention) when each *"command-item"* is *"Spotted"* (or *knowingly* *"Imagined"*).

However, let's do some practice with a seemingly unrelated *non-restimulative* exercise. For this, as you sit in a room (or *in-session*): *get a sense* (or *Imagine*) that each of the surrounding *walls* (in rotation) is telling you something. The statement said is: *"This means* ---"* (and then you fill-in various "words" to complete it).

Don't worry about what the *"this"* is; just practice with both standard (phrases used in everyday life) and ridiculous (fantastical) statements—but preferably not using the same one more than once in a row. This will give a *Seeker* an idea of the general "tone" or "style" of the *command-items* one might find in a pre-prepared *"implant sequence-list"* (used for *running* a specific *process*).

Let's do another practice concerning *"Spotting," "Imagining,"* and *"Confronting"*—this time using a *process* that might have a little more "bite" to it. There is no reason to target anything particularly *turbulent* for its practice—but, of course, as a *real process*, a *Seeker* should *run* whatever "comes up." By this point on the course, what we actually mean by *"spotting"* and *"confronting terminals"* should be well understood.

This is best practiced with eyes closed. You can either use *mental imagery* from *past memory, imagine/create* them as new, or even apply *ZU-Vision* (if such skills are available) to *actually see* whatever it is. This does not specifically target *implants* and is for practice only—but it can increase *Awareness* and contribute to *defragmentution* efforts, because it is actual *processing*.

Practice—Terminal: "People"

A1. *"Spot (Imagine) someone that you liked."*

A2. *"Look at them (and confront the imagery)."*

B1. *"Spot (Imagine) someone that you disliked."*

B2. *"Look at them (and confront the imagery)."*

Practice—Terminal: "Places"

A1. *"Spot (Imagine) a place that you liked."*

A2. *"Look at it (and confront the imagery)."*

B1. *"Spot (Imagine) a place that you disliked."*

B2. *"Look at it (and confront the imagery)."*

Practice—Terminal: "Objects"

A1. *"Spot (Imagine) an object that you liked."*

A2. *"Look at it (and confront the imagery)."*

B1. *"Spot (Imagine) an object that you disliked."*

B2. *"Look at it (and confront the imagery)."*

Practice—Motions: "Actions"

A1. *"Spot (Imagine) an activity that you liked."*

A2. *"Look at it (and confront the imagery)."*

B1. *"Spot (Imagine) an activity that you disliked."*

B2. *"Look at it (and confront the imagery)."*

Practice—Motions: "Events"

A1. *"Spot (Imagine) a time that you liked."*

A2. *"Look at it (and confront the imagery)."*

B1. *"Spot (Imagine) a time that you disliked."*

B2. *"Look at it (and confront the imagery)."*

When *implant-running*, a *Seeker* might sense that a certain *command-item* is actually "located" (as an *energy-mass*) somewhere near them, or perceives it being in a certain "direction." This may not occur; but if it does, a *Seeker* would *"mentally reach"* toward that location in order to *"Spot the Item."* Otherwise, the ideal intention is to *reach back* on one's own *"Spiritual Timeline"* and *Spot* "when" the *"item"* was *implanted.*

For *implant-defragmentation* to be at all effective, a *Seeker* must really "contact" or "connect with" an *implant item* in *processing.* There may be a sensation of "pressure" or a sense that you have contacted some *mass* or *energy*, similar to when you sense a *reaction* associated with some other kind of *"terminal"* or *"incident."*

The *"charge"* on (or attached to) a particular *"Item"* may not be very strong. Regardless, a *Seeker* repeatedly *"Spots"* it until there is literally *"nothing"* associated with it—*no* sensations of "heaviness" and certainly *no* feeling or desire to actually "obey" (or *"postulate"*) the *command-item* as one's own *consideration* or *creation.*

The same *Implant-Patterns* were used multiple times on the *Backtrack.* Similar to when *Confronting-the-Past* (or other methods of *incident-running*): once a *Seeker* has *"Spotted the Item"* multiple times, if there is no feeling of "relief" —and instead it seems to be getting "heavier" or more "turbulent" —it is likely that the same *command-item* was implanted during an earlier time (*implanting-incident*). Handling the *earliest/first* time is the only way to fully *defragment* any *implants* (or *imprints*) using *systematic processing.*

If there does not seem to be an earlier *implanting-incident*—and yet the *Item* seems to be getting more *"charged up"* — it is quite possible that some *charge/fragmentation* was left on one of the *earlier Items* in the same *sequence-list* (or *platform*) that you are presently *running*. This is easily remedied by looking back at the last few *Items* on your *list* to see if there is any *charge* still remaining there; and if so, just *run* that *Item* some more.

Implant-Running can sometimes require a lengthy *processing session* to complete. The intention is to take any *charge* or *fragmentation* off of an entire *sequence-list* (of *command-items*) that pertains to a particular *Implant-Platform*. In addition to a sense of *"relief"* (even if only minor), there should be no fatigue (or tiredness), sense of *mass* (or pressure), no *reactivity* (or significance), and especially no urge to "comply" with any of the *Items*. At that point, a *Seeker* can "casually" read over a *sequence-list* without concern.

Implanted command-items from a long time ago do not really have a lot of *influential* power once a *Seeker* is *Aware* of them. But they are what underlies all other *unknowing fragmentation effects* that accumulate (or stack up) as a *Being* continues their experience of existence onward from the *implanting-incident*.

For present purposes, we are mostly only concerned with handling those *basic implants* that *enforce* the *Human Condition* onto an individual. Other *Implant-Platforms* that *enforce* the fixed *reality-parameters* of this entire *Beta-Existence* (or *Universe*) in general, are of an even higher order of magnitude and are researched at much higher-levels of progress on the *Pathway*.

"IMPLANT-PLATFORMS"

In actual practice, once an *Implant-Platform* is fully brought into view — and *Known* — the residual *fragmentation* is fairly easy to *confront* or "shrug-off" if a *Seeker* has completed their previous *processing-levels.* "Layers" of *fragmented-energy-mass* are peeled off of these same "platforms" or "foundations" during standard *Beta-Defragmentation* procedures (given throughout the *Professional Course*)—and this is the only reason it is even possible for *implant-running* to be effective at these *upper-levels* of development.

Implants, almost by definition, get their "power" from being *hidden.* Even though a *Being* is "conscious" at the start-time of the event, an *implant* is "installed" or "attached" during an intense overwhelming and confusing *incident* that may actually include periods of "*unconsciousness*" (when an individual is literally "knocked out"). Much earlier on the *Backtrack,* before an *Alpha-Spirit* could be "hit" in this way, *implants* were often connected to "*aesthetically-pleasing*" ("*beautiful*") *incidents* that were simply "*hypnotic*" to look at.

Many of the *command-items* seem very extreme or "absolute" in their *wording* (*meaning*)—which generates *just* an intense enough impact to have *any* lasting impression at all. Most of the "negative" tone that is attached to an *Implant-Platform* is simply there to deter someone from *looking* at it; giving the sense that it would somehow be dangerous to remember, and therefore an individual avoids *mentally reaching* to that area in order to ever "*inspect*" it.

In many ways, this quality of "*unknowingness*"—and an individual's "*withdrawal from inspecting*" something—is true of *all fragmentation*; however, with an *implant,* the "loud bark" of its *systematic design* intentionally conceals just how "light of a bite" it actually has. Some of the more "extreme" *command-items* were designed to get an individual to "*postulate*" that they should forget the *implant* in order to *protect themselves.* But this *consideration* is made as a result of the *implant* and is not truly *Self-Determined*—and in actuality, the "harm" or "influence" of an *implant* will *lessen* the more it is *confronted As-It-Is.*

It cannot be too strongly emphasized that: the *implanted command-items* were never very powerful as a literal *command* or *control* "over" you. They are *creations* from an *external source.* The real *fragmentation* and *entrapment-to-agreements* occurred when *you* decided to make *"postulates"* and *"considerations"* either during the *implanting-incident* or immediately following it. Our own personal *fragmentation* (in an area) also multiplied exponentially whenever we were responsible for *implanting someone else.*

Systemology Society research resulted in pre-prepared *"sequence-lists"* for *running command-items* of specific *Implant-Platforms.* They are generally *run* at more *advanced processing-levels,* because even in *Traditional Piloting,* the *implants* are properly handled as *Solo-Processing.* To do this fully, a *Seeker* would already have had to have reached a basic state of *Beta-Defragmentation,* either from *Co-Piloting* or by making (at least) a second pass through all materials of the *Professional Course.*

Standard practices for *Solo-Processing* require a *Seeker* to isolate their attention on each *listed-item* individually. To accomplish this, you might place a sheet of paper over the *sequence-list* so that you can move it down to read each line; or cut an appropriately sized "rectangle" hole out of a piece of paper so that only one *command-item* is able to be viewed at a time. Then you simply *run* each *Item* as described in this lesson—and using your skills (and increased *Awareness*) from previous *processing-levels*—until there is *no* discomfort (*turbulence*) or *reactivity* of any kind while *considering* it afterward.

When *"processing-out"* an *Implant-Platform,* if you start to feel "really good," then take a break. This lets you have that "gain/win" before going back to the *list* and getting additional *charge* off it. After you've taken a break, if the *implant* seems resolved—humorous or ridiculous—scan the remaining *command-items* to see if there is any *charge* remaining (and then, of course, *run* what still does). Taking a break is an important part of the *process,* because sometimes there can be so much *"relief"* from a single *Item* that the whole *Implant-Platform* seems like it *defragmented,* when really it hasn't.

Rather than begin introducing entire *Implant-Platforms* at this late point in the lesson, a *Seeker* can practice *defragmenting command-items* with our specially prepared example (below). This *sequence-list* is not a complete *Platform*; however, our research demonstrated that it is often "added" to other *Platforms* as a deterrent to their discovery. Since we seldom attach this part to the actual *sequence-lists* (that are used for *processing*), we can provide it here, separately—and as a training example that provides actual gains.

IMPLANT DISCOVERY DETERRENT
COMMAND-ITEMS SEQUENCE-LIST

1A. *"To know about this is to disbelieve it."*

1B. *"To know about this is to forget it."*

1C. *"To know about this is to become insane."*

1D. *"To know about this is to become unconscious."*

1E. *"To know about this is to become less aware."*

1F. *"To know about this is to become sick."*

1G. *"To know about this is to die."*

2A. *"To talk about this is to disbelieve it."*

2B. *"To talk about this is to forget it."*

2C. *"To talk about this is to become insane."*

2D. *"To talk about this is to become unconscious."*

2E. *"To talk about this is to become less aware."*

2F. *"To talk about this is to become sick."*

2G. *"To talk about this is to die."*

3A. *"To remember this is to disbelieve it."*

3B. *"To remember this is to forget it."*

3C. *"To remember this is to become insane."*

3D. *"To remember this is to become unconscious."*

3E. *"To remember this is to become less aware."*

3F. *"To remember this is to become sick."*

3G. *"To remember this is to die."*

4A. *"To think about this is to disbelieve it."*

4B. *"To think about this is to forget it."*

4C. *"To think about this is to become insane."*

4D. *"To think about this is to become unconscious."*

4E. *"To think about this is to become less aware."*

4F. *"To think about this is to become sick."*

4G. *"To think about this is to die."*

After a *Seeker* *"processes-out"* all of the *command-items* of an *Implant-Platform*, the final (and most advanced) *defragmentation* step is to:

1. *"Spot any decisions (postulates and considerations) you made at the time of the implanting-incident."*
2. *"Spot any times you gave this implant to someone else (or wanted others to be implanted with it)."*

A *Seeker* is now prepared to move along further on the *Pathway*.

LESSON TWELVE
"GAMES AND UNIVERSES"

ENTRY INTO GAMES

In our previous lesson, we introduced *advanced incident-running* of *"spiritual implants"* —systematic suggestions from an external/outside (*other-determined*) *source* that have a tendency to affect the high-power *decisions* and *considerations* that we, ourselves, make about *"What-IS."* This brings us, then, to an upper-level point on the *Pathway* where the subjects of *"Games"* and *"Universes"* are handled more directly.

Participation in a *"game"* —whether it be the experience of this *Universe* in general, or another more specific activity involving *"choices"* —concerns the interaction between an individual (or individuals) and preset *"conditions."* The manner in which these *conditions* (or *aspects*) are handled (and *confronted*) is what distinguishes the experience of one *"player"* from another. It also gives some insight into what exactly an individual is attempting to accomplish—their *goals* and *purposes*—while *playing* the *"Game of Life."*

There is a *systematic sequence* (or *pattern*) to the way we are preoccupied with the *"roles"* and *"goals"* of our existence. This is quite helpful for retracing the unique steps of our *"spiritual descent"* on the *Backtrack.* But although we have uncovered *cyclic patterns* on the *Backtrack,* all individuals have not walked the same path *simultaneously* with each other. A *Seeker* is likely to be at a different *phase* of a *cycle* (or perhaps a different *cycle* altogether) than the next *Seeker*—just as those of us on the *Pathway* are all working at different paces.

Although we have tread into *advanced* territory, we will start off lightly with our *"Systemology of Games"* —first, by making certain a

Seeker understands what is meant by *"games."* We do not only mean a *game* like *"chess,"* although that *is* a *game.* We mean anything that involves an individual *playing a role* or *interacting with others*—especially when that *participation* involves working towards a particular *goal* or *purpose.*

Those *Seekers* with a large amount of experience with *systematic processing*—and additional *advanced ability levels*—have determined that *"Spotting"* the *"entry point"* of participation (when you *"entered"*) in a *game* (or even the *decision* to *participate*) can significantly *"release"* some of the *turbulent fragmentation* one might encounter when *processing* the *incidents* associated with that *game.* It is particularly effective if you can really *"Spot"* the point when your own interests and desires (*considerations*) drew you in to the *game.*

By itself, this practice does not automatically *defragment* all of the subsequent *incidents* that are connected. However, in these *upper-levels*: we are concerned with increasing greater *Actualized Awareness* in larger areas by *"unfixing"* (or *"freeing up"*) personal *"attention-units"* (personal *"spiritual energy"*) still *"stuck on"* (or suspended in) *"heavier"* *incidents.* By *confronting* these *incidents* as *"external"* from ourselves—or by seeing ourselves as *"exterior"* to the *incidents*—the magnitude of effect they have on us lessens enormously.

To start with: we will *consider* our involvement with others (*"relationships"* and *"groups"*) as *playing* a *game* or *role* in some way. We don't only mean *"sexual partnerships,"* where *relationships* are concerned, but any intense contact or involvement.

A *Seeker* might *"scan"* (or "look over") the events of a *relationship* to see if there is any *"charge"* or *"fragmentation"* present regarding a particular *incident*; but that is not really what we are treating at this level. We are most concerned with *"Spotting"* the *"entry points"* into any *relationships, group participation* (or *membership*), and *games.*

One might notice that happily established *"couples"* and *"partner-ships"* will often "reminisce" about their own shared *"entry-point"* into the *"relationship-game."* They frequently remind themselves (and each other) about *"how they got together"* in order to strengthen the "bonds" of the *relationship*—and to reinforce the "goals" that were (and are) "shared" for their own *"game."*

One might also notice that more *"turbulent"* and unpleasant *rela-tionships* consist of a lot of *"fluctuation"* in *attention* on the *"upsets"* that occur, rather than *shared-goals*. Some of these *"upsets"* stem from *misconceptions* that occurred near the beginning—if not at the actual *"entry point"*—of the *relationship*.

Rather than emphasize *mistakes*, a *Seeker* should focus on *"Spot-ting"* the original *desires* and *goals*—and the thing that you *thought was there*—while *"processing-out"* a *past-relationship*. Again, it is the *"entry-point"* into a *game* (or any *incident*)—the beginning, when it all started—that is most critical to *"spot"* when *processing*. This also takes unnecessary *"weight"* off of the things that happened later on in the *relationship*, which in turn makes any other related *defragmentation* or *incident-running* (given in earlier lessons) much easier.

Using what you have learned from earlier lessons (and tech-niques), if whatever you are *"processing-out"* seems to get *"heavier"* or more *"charged up,"* then you need to look for an "earlier begin-ning" (or *entry-point*) to that *incident* (or *game*); and if that isn't working, then look for an earlier similar *incident* (or *game*) that may be getting *restimulated* as a *"chain."* By this point, a *Seeker* will be experienced in handling these technicalities of *systematic pro-cessing*.

PARTICIPATION IN GROUPS

The next introductory area of *games* we treat is our participation in *"groups."* First, *"Scan"* the portion of the *Backtrack* (your past) that is accessible and list the various "*groups*" you *knowingly* (*willingly*) joined or were a part of (that you consider significant to your life-experience).

As described in the previous section, *run* the "*entry-point*" into each *group*. And as with "*relationships*," we are interested in the "*goals*," "*intentions*," and "*considerations*" that went into the *decisions* that led up to actually joining the *group*.

Another thing to pay special *attention* to—whether in joining *groups*, or handling *relationships* (previously)—is whether the *decision* was made as a *solution* to a previous "*problem*," or as a remedy for a period of "*confusion*." This is very important for *systematic processing*; because if that is the case, the "*decision*" to enter a *game* is actually in the "middle of the story," so to speak. Here, we would want to also "*Spot*" the *conditions* that existed prior to the "*decision*" —and of course, *what* that *decision* was attempting to *solve*. If this is case, *run*:

A. "*What problem (or upset) did you have?*"

B. "*What communication did you leave incomplete about that problem?*"

An alternative approach, assuming the "*problem*" was identified:

A. "*What did you do at that time?*"

B. "*What didn't you say at that time?*"

This should help take the "*confusion*" out of the "*chain*"—which allows a *Seeker* to handle the actual "*upsets*" or "*fragmentation*" that occurred (or was *restimulated*). The "*confusion*" part is what makes *defragmenting* the "residual effects" more difficult. The remainder is handled as described for "*Preventative Fundamentals*" in *Lesson-9*; which includes "*Flow-Factors*" (*Lesson-7*), "*Human Problems*" (*Lesson-4*), and "*Hold-Outs*" (*Lesson-6*).

There are also times when our participation in a *group* (or *relationship*, or any *game*) was "*enforced*"—in that we were "*forced*" to join or participate. Usually, our personal *protest* in such cases stems directly from the fact that we *perceive* the *"decision"* as *"other-de-termined"* from an external *source*, rather than *Self-Determined*. Our "power" of "free choice" has been reduced. To *"process-out"* such *fragmentation*, a *Seeker* would first *process* toward greater *"relief"* for the area of *"enforcement-to-join."*

A. *"Recall being forced to join something."*

B. *"What were you protesting then?"*

C. *"Recall forcing another to join something."*

D. *"What did they protest about it?"*

Once a *Seeker* has elevated their level of *Awareness* (and *ability-to-confront*) in this area, then they simply *"Spot"* any significant *entry-points* where their participation in a group was enforced, and then apply *incident-running* procedures (as learned throughout the *Professional Course*). It may be that all *"circuits"* (*you enforcing others; others enforcing others*) must be *run* fully in order to get the full *"release"* from this area.

BASIC "GOALS" AND "PURPOSES"

When an individual develops a strong *"purpose," "goal,"* or *"intention,"* it is an *entry-point* (or beginning, of sorts) into a *game*—because *goals* and *purposes*, by definition, are *"game-conditions."* In fact, a *game* is primarily distinguished by its specific *goals*: something an individual is trying to *do, have,* or *Be*. This is different from *"pure creation,"* because a *game* also consists of specific *"rules"* or *"parameters"* that define certain *"barriers"* and how a *"player"* is permitted to apply *effort* toward attaining the *goal*.

Sometimes a *goal* or *purpose* goes "unfulfilled" and is eventually "abandoned" altogether due to difficulties or failures in attaining it. These leave us with *fragmentation* of a *"failed purpose"*—whereas in a state of high-power *Awareness*, we might simply "set aside" one interest while pursuing another for the time being. As with other *entry-points* (described previously), *"Spotting"* the moment of the original *decision* (or *"postulate"*) greatly assists with *defragmenting* anything connected to it thereafter.

As with other areas *defragmented* for the *Pathway-to-Ascension*, our *processing* is not intended to enforce any *considerations* about what a *Seeker* ultimately decides to *do*, or *participate in*, or *have*, *&tc*. The purpose of *defragmentation* is to rehabilitate the original power of *"free choice"* that an *Alpha-Spirit* experienced before all of their *attention* or *Awareness* became confined to the most restrictive of *considerations* (and *"postulates"*) for the level of *game* taking place in *this Physical Universe*.

Once *fragmentation* is eliminated from a *"failed purpose,"* the basic desire behind the underlying *goal* may be *revitalized*, or the *Seeker* will see it *"As-It-Is"* and lay it to rest depending on what their *actual goals* are for the present. These older ones (considered *"failed"*) tend to get in the way of our clear handling of present goals—mainly due to the amount of *attention-energy* still suspended *unknowingly* in that area. When this *attention* is freed up, you can more clearly *consider* all of the *goals* you could have.

A. *"Spot a goal (or purpose) desirable to you."*

B. *"Spot a goal (or purpose) desirable to someone else."*

As an additional step further—for *advanced Seekers*—on future passes through the *Professional Course*, it is very beneficial to *"Spot"* and *run* the *incident* of "picking up" your current *Body*. This is perhaps one of the most critical (and earliest) *entry-points* an individual has concerning the *Game* of *this* lifetime. Of course, we have picked up many bodies, and have had repeated *incidents* of *entering into this Universe* before; hence this area requires *systematic processing* by a very experienced practitioner.

ENTRY INTO UNIVERSES

It takes a high-level of *Actualized Awareness* to fully understand and appreciate the phenomenon we refer to as the *"Condensation of Universes."* Many esoteric and spiritual traditions have produced various mystical models of *"interdimensional trees," "stargates"* and *"kabbalahs"* to demonstrate this—but too often the true intended meaning has been lost to thousands of years of reinterpretation and cross-communication.

An *entry-point* into *this Universe* (or version of *Beta-Existence*) differs from the *entry-point* into an *earth-bound* experience of the *Human Condition*; although, as we've described (above), that too is an *entry-point*. There are obviously many levels of *"game"* taking place simultaneously—and part of *upper-level defragmentation* involves *"Spotting"* and *"differentiating"* between the various *games* (and their *"parts"*) that we are still participating in, *knowingly* or otherwise.

Of course, as an *Alpha-Spirit* we are essentially *"non-local,"* but as we have covered in other lessons: the *condensation* (*compactification* and *solidification*) of our total *Awareness* as *Self* has become fixed to (or *"collapsed-in"* on) one *singular viewpoint*—and in our case, most recently on the *Backtrack*, fixed to a location *"interior to"* the *Human Condition*, on *Planet Earth*, in *this Physical Universe*.

Our true *Beingness* is not *actually located* behind some *"eyes"* or in a *"head,"* but for all intents and purposes, this is how we have been accustomed to operate as a *"player"* in the more *"localized"* game of *Beta-Existence*. At some point on the *Backtrack*, an *Alpha-Spirit* became dissatisfied with operating a *game* while *exterior-to* it, and *"decided"* that a more substantial *"immersive first-person"* level of experience was the solution. This was later added to with a reinforcement of *"pleasure"* and *"pain"*—and to take things a step further and worse: eventually an individual *"decided"* to *forget* they were *playing a game* altogether.

Throughout our *Systemology*, we often draw a distinction about "*this*" *Universe* or "*this*" version of *Beta-Existence*—which just summons the mind to question what "*other*" *Universes* or versions of *Beta-Existence* we might otherwise be referring to. [If at any point this all seems "*too speculative*" for a *Seeker*, return to this subject on your second pass through the *Professional Course.*]

Of course, we have previously mentioned (in former lessons) the "*Personal Universe*" (or *Home Universe*) where the *actual Beingness* of the *Alpha-Spirit* still remains, and of which the experience of *Beta-Existence* is "superimposed" over. But *this* present version of *Beta-Existence* is not the *only Universe* that has ever been "superimposed" and experienced, as new data from the *Backtrack* consistently reflects and confirms.

We covered a basic understanding of *implants* in the previous lesson. The *entry-point* to *this Universe* also includes an "*implanting-incident*" in which to "transition" an individual's *Awareness* (and sense of *Beingness*) *out of* the "previous" *Universe*. The two *Universes* are not otherwise spatially connected in any way. You could not, for example, travel any distance in *this Universe* and somehow find yourself in the former one.

The only "bridge" between *Universes* is the "*implanting-incident*" itself. And the entire *implanting-incident* is manufactured (or *created*), *existing* within its own *space-time* just as any other *Universe*. Its range of potential experience is much "smaller" relatively (compared to *this Universe*, for example), since it operates on a "*prerecorded loop*" of sorts; but it is obviously quite efficient in "*implanting*" our perceived parameters for the level of *game* taking place "down here" in *this Universe*.

In order to be *systematically concise* in our work, we often draw a distinction between *Universes* by referring to *this* version of *Beta-Existence* as "*this version*" or the "*Physical Universe*." By relative contrast, the previous or former *Beta-Existence/Universe* was quite "*magical*"—and hence we distinguish it with the title "*Magic Universe*" or "*Magic Kingdom Universe*." It is not simply some other "planet" within *this Universe*, or even an obscure *"astral"* dimension *"in-between" space*, or anything like that; it is an entirely separate *Universe*.

The entire sequence concerning our *Entry-into-this-Universe* really has an "earlier beginning" in the *Magic Universe*. The *transition point from* the *Magic Universe* is often described as being "pushed in to" a surrealistic (usually outdoor) pool, backed by *Greco-Roman*-style pillars and a beautiful sky. We could outline the full sequence of events briefly as this:

1. *Magic Universe* (previous *"Beta-Existence"*)
2. A *transition* point *from* the *Magic Universe*
3. A *"mini-verse"* bridge (recorded *implanted-incident* between *Universes*)
4. *Physical Universe* (entering this *"Beta-Existence"*)

Unlike some former more "recreational" existences: at its inception, this *Physical Universe* was *created* as a place to penalize, exile, or otherwise imprison, the criminals and maladjusted individuals from the *Magic Universe*. This paints a much different picture of things than the fluffy politically-correct idea impressed by some *mystics* and *spiritual leaders* regarding this existence being some kind of *"school."* About the only thing an *Alpha-Spirit* has of value to *learn* "down here" is *why and how they got here* and *what is the best way out.*

Of course, there are some who *have* "escaped" in the past—and so there may be repeated *incidents* of *entry-into-this-Universe* (and as usual, one looks to *spot* the earliest one for *defragmentation*). But, this has no longer been the standard *goal* and *purpose* of a *Being* entrapped down here. For one: an individual no longer carries a clear *Knowingness* of their "entrapped" circumstances—so fewer attempts and efforts are even made.

There is another factor that has taken place as well: as more of the population from the *Magic Universe* ended up in the *Physical Uni-*

verse, there were fewer remaining to maintain the *"creation"* of the *Magic Universe*. Things eventually became *"more interesting"* in the *Physical Universe*, and many within the *Magic Universe* migrated down here *knowingly* and by choice. Such individuals would at first have appeared to be quite *"advanced"* (or like *"gods"*) in contrast to the "more primitive" individuals entrapped in *Beta-Existence*.

The *transition-point* (involving the *"pool"*) is a significant experience to *"Spot"* when handling the *incident*. Unlike certain factors that are common to everyone that has the experience, the actual "reason" for such a "legal sentencing" is likely to be somewhat unique for each occurrence. [More data is still need in this area.]

For whatever the reason, an individual finds themselves being "drawn down" to descend a *"spiral of pillars"* into the *pool*. Whatever *Beings* are present (above/around the *pool*) also use their *energy* to "push down" upon the individual as well—thereby participating in the eventual *"implanting-incident"* that ensues. Once the transfer occurs, the *implanting-incident*, from this point onward, is always the same prerecorded pageantry.

THE IMPLANTING INCIDENT

Data for this *"implanting-incident"* accumulated over several years. It is consolidated from exploratory research from many *"advanced practitioners"* (working beyond *Beta-Defragmentation*) using *GSR-biofeedback devices* (and subsequently comparing notes afterward). As such, there are some details that may be in error (or absent), but enough valid information was recovered for a *Seeker* to *"Spot"* and *run* the *incident*. To start, let's simply look over the data.

This *implanting-incident* does not contain much "pain" (or "force"). It is a *"mini-verse"* manufactured in the *Magic Universe* to be *"aesthetically-beautiful"* so as to hold one's *attention* in and eventually draw the *Alpha-Spirit* into the *Physical Universe*. Any *"turbulence"* connected to the *incident* itself, is merely a sense of *"loss"* (over leaving the *Magic Universe*) or misemotion regarding *"exile"* (or being *"pushed out"*).

There are really two parts to this *incident*: the *first* sets up the *second*. In the *first* part: an individual finds themselves essentially floating in a *"void-like space"* with a sensation similar to being *"under water."* They eventually see a *"golden light"* in the distance—it glimmers and radiates like a sunburst, but more like a reflective piece of jewelry rather than an actual light-source itself.

Once the individual's *attention* is fixed upon it, the *implanting* begins. They are "drawn toward" it with increasing speed. This *"golden light"* is an *"Object-Item"* that is attached to the first *"Implanted Goal"* of an entire sequence (consisting of 36 *"Goals"* or *"archetypes"*). [In this case: the first one, is *"To Be Godlike."*]

Each *"Goal"* is introduced and identified with a particular *"Object-Item"* (in order to communicate and embed its significance). The *"Goals"* *are* named, but the titles are *sensed* (or *intuited*) rather than expressed in words; we can only *approximate* their full meaning with *Human* language (for *study* and *processing*).

The individual is "drawn in and through" each *"Object-Item"* (they are apparently holographic)—and while passing through, the *"Goal"* (and its label) is identified and understood. This continues with increasing speed through the remaining sequence— simply informing *what* the *"Goals"* *are*, in order to set up the *second* part of the *incident*, when this *"Goals-Sequence"* is *implanted* again with *significance*.

In the *second* part: the *Being* emerges from the final *"Object-Item-Goal"* in this sequence (a *"pyramid"* associated with *"To Be Enduring"*) onto a different and more substantial plane or landscape that involves an "amphitheater" and "stage."

Some *Advanced Seekers* have referred to this whole *incident* as the

"*Heaven Implant*," because the scenery (of this *second* part) reflects a stereotypical angelic "*trumpet-blasting-cherubs-in-the-clouds*" atmosphere that *Humans* are *implanted* to "expect" for their *after life* (or more correctly: "*between-lives*" period). [In actual fact, the real "*Kingdom of Heaven*" (such as *Jesus* speaks of) is the *Magic Kingdom Universe*, and not this *implanting-incident* that everyone has been forced to experience (multiple times).]

The *second* part of the *incident* mainly entails a "theatrical pageant" or "skit." There is the appearance of other individuals in attendance; but it is unlikely that you would experience this simultaneously with other actual entities (*Alpha-Spirits*), so this crowd that gathers is probably fabricated as part of the recorded scene. Sometimes there is a feeling of being overwhelmed by the surrounding crowd as everyone excitedly pushes up toward the stage.

There is a trumpet-horn that blasts, and a sharp snapping-crackle (like the pierce of thunder), to get your *attention* right before any "*command-line*" is given. The first *command-line* is from an unseen *source* (to get the "pageant" started)—"*Only One Will Survive.*" Again, the horn blasts, the snaps happen, and the second *command-line* emerges—"*To Be The One Who Survives, You Must Be Superior To All Others.*" This seems to settle the crowd down and all *attention* becomes focused on the stage.

The blasts and snaps are also heard as a new "character" (each representing one of the "*Goals*") comes on the stage to speak *three* "*command-lines.*" Only one "character" is on stage at a time. They always enter from one side of the stage (your right) and exit the other (your left). There is a *procession* or *sequence* taking place, so the "characters" tend to *look* in a particular direction when referring to the *next* (or the *previous*) "character."

When the *final* "character" (*An Enduring Being*) completes their reference to the *first* (*A Godlike Being*), the *implanting-incident* ends with "waves of blackness" before the individual finds themselves as an *Alpha-Spirit* with fixed *viewpoints* in *this Physical Universe*

(*Beta-Existence*). And while there are many *entry-points* in this *local Universe*, according to research, the most recently used one is the *Horsehead Nebula* (*Orion*).

The *110 "command-items"* of this *Implant-Platform* are given in the next section (below); and the significance for each is *"processed out"* (as described in *Lesson-11* *"Spiritual Implants"*). The re-searched description (above) simply helps in *"Spotting"* and *"connecting with"* (contacting) each *Item* on the *Backtrack*. But before doing any *processing*, let's examine exactly what is taking place with this *Implant*.

The *incident* is *implanting* a *game*. The emphasis is *"survival"* rather than *"creation."* It is intended to sell you on the idea that the *game* of *this Universe* is *"survival by superiority"* —that you must be *"superior"* to everyone else, and only one will *"survive"* —which immediately puts you into *conflict* with everyone else and auto-matically sets you up for failure. And of course, there is no actual *"winning"* of the *Game*. It simply occupies *attention* while making its *Players* just a little bit worse off with each *"lifetime."*

When you're "fighting against" *everyone* else, it is only a matter of time before you *"lose."* While everyone is originally *implanted* to operate from a "godlike" state at the "start" of the *game*, no one is ever actually able to maintain that state (in opposition to everyone else); But as we "fail" with each "*Goal*," our perceived "*purpose*" sinks down through the remaining sequence of "*Goals*" —until we end up at "*To Be Godlike*" again, but this time a "lesser version" of what we were before, as we continue down a dwindling spiral.

The entire "*Implant-Platform*" is a "lie" —built upon the foundation of the very first and most significant lie that "*only one will survive.*" This *implant* is the primary reason for conflict in this *Game-Universe*; the entire foundation of *this existence* is essentially "bullshit."

When a *Player* finally calls "bullshit" on the *Game*, and can fully stop using their *skills* and *abilities* against others (and for *superiority*), then they can break free of the "game-pattern" that most strongly fixates *considerations* and *viewpoints* to *this Universe*. Does *"processing-out"* the following *Implant-Platform* automatically release you from *Beta-Existence* and return you to the *Magic Universe?* Certainly not. But, if that *is* a direction you are intending to go, than this type of *implant-running* is a very critical step.

THE GOAL-SEQUENCING IMPLANT

The following *110 Command-Items* compose the basic *Implant-Platform* for *"Entry-to-this-Universe"* (and entry into *this Game*). Any personal significance attached to these *Items* is *processed-out* (using the training and procedure given in *Lesson-11*). While *processing*, try to *"Spot"* the *Items* *"in"* the *incident*, using the details described in the previous section (*imagining* when necessary).

0A. *"Only one will survive."*

0B. *"To be the one who survives, you must be superior to all others."*

1A. *"To be godlike is to solve the opposition of enduring (or stubborn) people."*

1B. *"To be godlike is to be superior to all others."*

1C. *"To be godlike is to suffer the oppression of free beings."*

2A. *"To be free is to solve the opposition of godlike beings."*

2B. *"To be free is to be superior to all others."*

2C. *"To be free is to suffer the oppression of responsible beings."*

3A. *"To be responsible is to solve the opposition of free beings."*

3B. *"To be responsible is to be superior to all others."*

3C. *"To be responsible is to suffer the oppression of creative beings."*

4A. *"To be creative is to solve the opposition of responsible beings."*

4B. *"To be creative is to be superior to all others."*

4C. *"To be creative is to suffer the oppression of important beings."*

5A. *"To be important is to solve the opposition of creative beings."*

5B. *"To be important is to be superior to all others."*

5C. *"To be important is to suffer the oppression of competent beings."*

6A. *"To be competent is to solve the opposition of important beings."*

6B. *"To be competent is to be superior to all others."*

6C. *"To be competent is to suffer the oppression of famous beings."*

7A. *"To be famous is to solve the opposition of competent beings."*

7B. *"To be famous is to be superior to all others."*

7C. *"To be famous is to suffer the oppression of perceptive beings."*

8A. *"To be perceptive is to solve the opposition of famous beings."*

8B. *"To be perceptive is to be superior to all others."*

8C. *"To be perceptive is to suffer the oppression of energetic beings."*

9A. *"To be energetic is to solve the opposition of perceptive beings."*

9B. *"To be energetic is to be superior to all others."*

9C. *"To be energetic is to suffer the oppression of meticulous beings."*

10A. *"To be meticulous is to solve the opposition of energetic beings."*

10B. *"To be meticulous is to be superior to all others."*

10C. *"To be meticulous is to suffer the oppression of successful beings."*

11A. *"To be successful is to solve the opposition of meticulous beings."*

11B. *"To be successful is to be superior to all others."*

11C. *"To be successful is to suffer the oppression of accurate beings."*

12A. *"To be right (accurate) is to solve the opposition of successful beings."*

12B. *"To be right (accurate) is to be superior to all others."*

12C. *"To be right (accurate) is to suffer the oppression of popular beings."*

13A. *"To be popular is to solve the opposition of accurate beings."*

13B. *"To be popular is to be superior to all others."*

13C. *"To be popular is to suffer the oppression of skillful beings."*

14A. *"To be skillful is to solve the opposition of popular beings."*

14B. *"To be skillful is to be superior to all others."*

14C. *"To be skillful is to suffer the oppression of wise beings."*

15A. *"To be wise is to solve the opposition of skillful beings."*

15B. *"To be wise is to be superior to all others."*

15C. *"To be wise is to suffer the oppression of beautiful beings."*

16A. *"To be beautiful is to solve the opposition of wise beings."*

16B. *"To be beautiful is to be superior to all others."*

16C. *"To be beautiful is to suffer the oppression of productive beings."*

17A. *"To be productive is to solve the opposition of beautiful beings."*

17B. *"To be productive is to be superior to all others."*

17C. *"To be productive is to suffer the oppression of powerful beings."*

18A. *"To be powerful is to solve the opposition of productive beings."*

18B. *"To be powerful is to be superior to all others."*

18C. *"To be powerful is to suffer the oppression of holy beings."*

19A. *"To be holy is to solve the opposition of powerful beings."*

19B. *"To be holy is to be superior to all others."*

19C. *"To be holy is to suffer the oppression of intellectual beings."*

20A. *"To be intelligent is to solve the opposition of holy beings."*

20B. *"To be intelligent is to be superior to all others."*

20C. *"To be intelligent is to suffer the oppression of strong beings."*

21A. *"To be strong is to solve the opposition of intellectual beings."*

21B. *"To be strong is to be superior to all others."*

21C. *"To be strong is to suffer the oppression of crafty beings."*

22A. *"To be crafty is to solve the opposition of strong beings."*

22B. *"To be crafty is to be superior to all others."*

22C. *"To be crafty is to suffer the oppression of brave beings."*

23A. *"To be brave is to solve the opposition of crafty beings."*

23B. *"To be brave is to be superior to all others."*

23C. *"To be brave is to suffer the oppression of wealthy beings."*

24A. *"To be wealthy is to solve the opposition of brave beings."*

24B. *"To be wealthy is to be superior to all others."*

24C. *"To be wealthy is to suffer the oppression of independent beings."*

25A. *"To be independent is to solve the opposition of wealthy beings."*

25B. *"To be independent is to be superior to all others."*

25C. *"To be independent is to suffer the oppression of morally good beings."*

26A. *"To be good is to solve the opposition of independent beings."*

26B. *"To be good is to be superior to all others."*

26C. *"To be good is to suffer the oppression of adventurous beings."*

27A. *"To be adventurous is to solve the opposition of good beings."*

27B. *"To be adventurous is to be superior to all others."*

27C. *"To be adventurous is to suffer the oppression of orderly (organized) beings."*

28A. *"To be orderly is to solve the opposition of adventurous beings."*

28B. *"To be orderly is to be superior to all others."*

28C. *"To be orderly is to suffer the oppression of different (eccentric) beings."*

29A. *"To be different is to solve the opposition of orderly (organized) beings."*

29B. *"To be different is to be superior to all others."*

29C. *"To be different is to suffer the oppression of respected beings."*

30A. *"To be respected is to solve the opposition of different (eccentric) beings."*

30B. *"To be respected is to be superior to all others."*

30C. *"To be respected is to suffer the oppression of happy beings."*

31A. *"To be happy is to solve the opposition of respected beings."*

31B. *"To be happy is to be superior to all others."*

31C. *"To be happy is to suffer the oppression of acquisitive beings."*

32A. *"To be acquisitive is to solve the opposition of respected beings."*

32B. *"To be acquisitive is to be superior to all others."*

32C. *"To be acquisitive is to suffer the oppression of sensual beings."*

33A. *"To be sensual is to solve the opposition of acquisitive beings."*

33B. *"To be sensual is to be superior to all others."*

33C. *"To be sensual is to suffer the oppression of domineering beings."*

34A. *"To be domineering is to solve the opposition of sensual beings."*

34B. *"To be domineering is to be superior to all others."*

34C. *"To be domineering is to suffer the oppression of tough beings."*

35A. *"To be tough is to solve the opposition of domineering beings."*

35B. *"To be tough is to be superior to all others."*

35C. *"To be tough is to suffer the oppression of enduring (or stubborn) beings."*

36A. *"To be enduring is to solve the opposition of domineering beings."*

36B. *"To be enduring is to be superior to all others."*

36C. *"To be enduring is to suffer the oppression of godlike beings."*

GAMES AND UNIVERSES

The subject area of *"Games and Universes"* is quite advanced, even for our *Systemology*. In essence, a *Seeker* is *indirectly* treating this same area with *systematic processing* the whole while they are progressing on the *Pathway*—right from the beginning of *Systemology Level-0*. But, when it comes to *really understanding* what is happening with us and all around us, it essentially boils down to *"Games and Universes."*

It should be evident from this lesson that *this present Game-Universe* is based on *"force"* and *"conflict"*—which are used toward achieving a *"superiority"* that never actually arrives. This is not so surprising, given the original *intention* for this *Beta-Existence* as primarily a *"spirit-prison."* But, this is not the first *Game-Universe* that developed as a *"penalty"* of sorts. Prior to this, the *Magic Universe* also once began as a *secondary* universe to the one that preceded it. So these *"cycles"* have been taking place for a very long time.

Game-Universes aren't *"new"*; they actually start to appear quite early on the *Backtrack*. And *Alpha-Spirits* as *"Eternal Beings"* really like to play *"games"*—to *have* something to *do* in order to *Be*. This is the reverse of what our original native state is, where we can simply *Be* anything—and even *create* anything—*at will*.

But, as is even evident by the popularity of certain types of "*video games*" today, the *Alpha-Spirit* likes to "*play*" when it is not engaged in "*creating*" things to play with. And, of course, this has gotten us into a bit of trouble along the long course of our *Spiritual Existence*. The *Alpha-Spirit* eventually lost the *willingness* and *ability* to *confront* the *Infinity-of-Nothingness*, and now prefers literally *any game* over that native state.

In the earliest *Shared-Universes*, *games* were handled with much more primitive novelty—not very different from modern "*motion pictures*," "*video games*," and even "*virtual reality*." They were conducted *knowingly* for *entertainment*; a way to interactively "show off" our *creations* to one another. They were never meant to reach such a level as to *unknowingly entrap spirits* or employ *actual living things to suffer as playing pieces*—and yet this is the real truth about how far we have descended in our existence as *Spiritual Beings*.

In order to progress beyond the upper-most reaches of the *Pathway*, a *Seeker* will have to eventually regain the use of *force* and *energy* (from a perspective as an *Alpha-Spirit*, not a *Human*). This area of development is only emphasized *after* a *Seeker* can be certain that they are mastering it to regain *control* of the *conditions* that have *entrapped* them, and not to use it to further the *domination/superiority-game* with others. [Doing so is technically how we *lost* or *forgot* these *spiritual abilities* in the first place! If you *knowingly* continue that *game* after increasing *Awareness* and personal power on this course, you will likely sink so fast, like in quicksand, that even *we* might not be able to pull you back out again.]

There is nothing inherently wrong with *games*. There is nothing even inherently wrong with the "*Goals*" of *this Universe*, themselves. Our own goal in *defragmenting "The Goals"* is for a *Seeker* to retain the "positive" characteristics that each suggests, without being compelled to adopt the "inverted" *considerations* that they should be "weaponized" or used against others (which is what the *Implant* is really *impressing*).

By following along with the *Implant* (*Sequence*) we encounter unnecessary invisible *barriers* that restrict our very *considerations* of *Beingness*. For example: the *Implant* suggests that we should have issues *being* both *strong* and *intelligent*, or with *being responsible* and a *Free Spirit*, *&tc*. While the "*command-items*" do not literally say these things, these are the type of *Alpha-Thoughts* ("*postulates*") that an individual will *consider* on their own for themselves purely as a result of having experienced the *Implant*.

This whole area of "*Implanted-Goals*" is very closely associated with "*Justification Considerations*" (see *Lesson-8*) that *Seekers* uncovered about themselves previously to complete *Systemology Level-3*. This is where we did a bit of hunting to find out "*what makes one superior*" and "*what makes others wrong*." The answers to this will often provide a clue as to where on the "*Goal-Sequencing cycle*" a *Seeker* is likely to be. In other words: *Justification Considerations* usually indicate which *Implanted-Goal* a *Seeker* is presently occupied with.

One exercise that can help free up these *considerations* is: walk around and *get the sense* (or *Imagine*) yourself as having the primary positive characteristic of a "*Goal*." Then *look* around and *Spot* different people, *Imagining* that each of them also carries that strong positive characteristic (even if they are not visibly demonstrating it). Now, *Imagine* that there is *more* of that characteristic — being *more godlike*, or *stronger*, *&tc*. Alternate: putting the *intention* into yourself, and putting the *intention* into others, [Keep in mind that we have been "competitively programmed" to actually maintain the opposite *intentions* about others.]

There are some *advanced processes* concerning *Universes* that are used as a "checkpoint" for *Seekers*, here at the completion of *Systemology Level-5*. They may be used to increase "*spiritual perception*," but for our present purposes, they are intended to "unstick" or "unfix" a Seeker's attention and Awareness from being so compulsively on this Game-Universe (Beta-Existence). The first exercise is adapted from the original "Wizard Training Regimen" developed by the Systemology Society many years ago. The second/final process is an advanced application of a familiar technique we've used previously in the Professional Course. These are "creativeness processing" exercises that may be run using one's "imagination" (or "visualization") until a Seeker gets a sense that their spiritual perception has increased.

 * EXERCISE #1 *

Sitting comfortably, eyes closed.

• Imagine you are extending your Awareness, reaching through the entire Physical Universe (as far as you can imagine). Now reach a bit further beyond all perception of dimensional space until you find "Nothing." Hold your attention (point-of-view) on the "Nothingness" without thinking of, or imagining, anything else.

• As before: specifically extend your reach out on the right side; contact and hold on the "Nothingness."

• Practice as before: extending your reach out to each direction (by itself); the left side, in-front, behind, above, and below.

• Now extend your reach out in two directions (simultaneously); making sure to contact and get a sense of the "Nothingness" on both sides (from your point-of-view)—right and left; then in-front and behind; and finally, above and below.

• When a Seeker is well practiced in the above: extend your reach in all six directions simultaneously, getting a full certainty of the "Nothingness" on all sides.

 * EXERCISE #2 *

Laying down, comfortably, eyes closed.

Alternate.

 A. "Spot three points in this Universe."

 B. "Spot three points that are not in this Universe."

This completes Systemology Level-5.

LESSON THIRTEEN
"SPIRITUAL ENERGY"

INTRODUCING "ENERGY"

The remaining lessons of the *Professional Course* pertain to *Systemology Level-6*, which means that we are now treating areas officially considered *"Advanced Training"* or *"Ability Training."* Upper-levels are usually only introduced *after* a *Seeker* reaches a basic state of *"Beta-Defragmentation"* using previous *processing-levels* (*Systemology Level-0* to 5); and this requires a *Solo-Pilot* to make multiple passes through the material.

We introduce *Systemology Level-6* during the *Professional Course* to provide a greater sense of "completion" to the material we have covered—and to extend an invitation for *Seekers* to continue their progress with the *Advanced Training Course* that will later follow (*scheduled for publication in 2024*). Upon reaching a basic state of *Beta-Defragmentation*, a *Seeker* is quite aware that there is much more of the *Pathway* ahead to still *reach* for; and they are quite ready and interested to *see what's next*. And our answer is: *Systemology Level-6*.

This lesson introduces subjects that a *Seeker* may be familiar with —by label—from other *mystical* or *magical* materials and/or commonplace pop-culture representations of the *"New Age."* By this we mean other times *"Spiritual Energy"* is taught in terms of *chakras, auras, etheric bodies*, and the like. A *Seeker*/student is advised to study and practice these *metaphysical* areas *only within our Systemology* while on this course; and not presume to "know all about it," or mix previous interpretations with *this systematic methodology*.

Since the beginning of the *Pathway*, a *Seeker* is essentially dealing with *"energy"* in *systematic processing*. When we speak of *attention*, *circuits, flows, imprints* and *mental creations*, we are actually treating *"energy"* indirectly. The entire *applied philosophy* of *Systemology* is in many ways entirely about *"energy"*—but we do not introduce any exercises aimed at directly *"handling energy"* until a *Seeker* has reached these *upper-levels*.

The *"spiritual weight"* of *fragmentation* entangled in the areas previously covered on this course is not all that *"holds us"* in *this Physical Universe* (*Beta-Existence*). To truly *"Ascend"* and be free of *this Beta-Existence is* to actually break the *"gravity"* or *"pull"* that this *reality* has on the *Awareness* of our *Beingness*. A *Seeker* will eventually have to be able to fully *confront, handle* and even *create* *"energy"* as an *Alpha-Spirit*, in order to reach the *"ultimate"* end of the *Pathway*.

The foundations of *this Beta-Existence* are based entirely on the use of *"force"* in order to maintain the level of *"solidity"* we experience *here* everyday. By *"force"* we mean *"energy* with *direction."* At the most physical level of its manifestation, we might equate this with *"effort."* But *Spiritual Energy*—or *"ZU"* (to revive a *6,000+* year old word for it)—is not limited only to expressions of *"effort"* demonstrated by *physical genetic-vehicles* or *bodies*. We are not restricted to only *handling energy* as it is expressed visibly in *this Universe* at all.

Beta-Existence is entirely composed of *energy*; but it is a type of *energy* that only operates in a *"space"* called: *"This Physical Universe."* An *Alpha-Spirit* that is convinced it *must rely* on *Beta-Existence* for its *energy* will never give up its own *"hold on"* it. There is a belief that all potential *energy-matter* of *all existences* is *"fixed"* (or *"conserved,"* as the standard physics theory is labeled)—limited to what is already provided for us, by some *other-determined source*, to *change* with *effort*. This falsehood is a large part of what keeps our *Awareness* of *Beingness* entrapped here.

Along the *track* of an individual's *spiritual descent* into the *reality-agreements* of *this Universe* and *Human Condition*, there is an incremental decline in the *Alpha-Spirit* maintaining itself as "*cause*" or "*source*." Rather than *cause* its own *effects* for-and-as *Self*, the *Being* decides on more and more "external" things to be *cause*, so that they may receive the *effects* of, for example, a *perception* or *sensation*. This establishes a very *real* (though *artificial* in *actuality*) "dependency" on the *Physical Universe* and "*Bodies*" in order to experience existence. It becomes practically a "spiritual addiction" that we typically can't shake off.

A complete mastery of *handling energy* exceeds the scope of this lesson. However, there are many *systematic processes* and exercises that help a *Seeker* develop more in specific areas, while at the same time increasing their general "*Actualized Awareness*" —the overall "*Spiritual Power*" that the *Seeker* eventually must use to get free of this *Beta-Existence*.

BASIC CHARACTERISTICS OF "ENERGY"

Seekers more easily understand the true characteristics of *energy* after practicing the procedures given for earlier *processing-levels*. This is because demonstrations of *energetic characteristics* are experienced during the *processes*, even when they are not directly and distinctly identified as such. Before getting too deep into this area, we'll introduce the idea of "*energy*" with light exercises you can practice both indoors and outside.

A1. "*Spot an energy that could be helpful to you.*"

A2. "*How could it be helpful?*"

B1. "*Spot an energy that could be helpful to others.*"

B2. "*How could it be helpful?*"

A *Seeker* can "*Spot*" energies that are physically "visible" (such as a "light" that is "on"), or mentally "*Spot*" something apparent but hidden (such as the "current" in the walls feeding the "light"). There are no right or wrong answers here. It simply prepares a *Seeker* to *run* the following *process in-session*, listing as many *sources* as come up—whether *"Spotted"* as *real* in this *Beta-Existence*, or "conceptually" in another *Universe*.

1. *"Spot an energy source you find acceptable."*
2. *"Spot an energy source others would find acceptable.*

Control of a "system" consists primarily of the *start, stop,* or *change* of some *energetic-flow*. *Energy* simply *is* "energy." But when we *do* something *with* it, or *to* it, we are affecting its "*characteristics*" —or else making it the "type" of *energy* it is, or rather, how it is experienced.

A *Seeker* can physically practice *getting a sense* of *controlling* the "start" and "stop" of an *energy-flow* by, for example, using a functioning "light-switch" or simple "electronic device." Start with the switch in the "OFF" position. Alternate these *processing command-lines* ("PCL") a few times, or until you feel good about your actual perception of the *energy-flow*, even if only vaguely sensing it. [A similar *New Thought* exercise uses a "water-faucet" to practice *controlling* a "*flow*."]

A. *"Mentally reach into the device (or circuit), permeate it, and perceive the 'no-flow' of energy."*
B. *"Continue to mentally permeate it as you (physically) turn it on, and perceive the 'flow' of energy present."*

Handling energy is not restricted to manifestations found in the *Physical Universe*. In the "*Magic Universe*" that precedes (or is a *level higher* than) *this Universe*, use of what *Humans* call the "electron" (as a *practical energy*), is *not* restricted to *flows* on "wired-circuits," or *consuming* a "limited resource" in order to *generate* it. There is still a use of focused "*energy*," but by relative comparison to this *Beta-Existence*, experiences there really would seem much more "*magical*."

When a *Seeker* "mentally" handled *space* using "*dimension-points*" and "*corner-points*" (or *anchor-points*) in some exercises from earlier lessons, they were treating the most basic unit of *energy*. When *energy* is in *action* it takes on specific *characteristics*.

A *"flow"* is the characteristic of *energy* being *transferred* or *communicated*. For example: in *systematic processing*, we *run* "circuits" that represent specific *energy-flows*; such as an *"out-flow"* from you to another *"terminal"* (lifeform, location, mass, *&tc.*), or an *"in-flow"* from a *terminal* to you. These are distinctions of the primary *flow-direction* of an *energy-beam*.

Direct practice with *energy-beams* is reserved for *upper-levels*, because if a *Seeker* starts working with them by trying to *do* something with them (to affect a visible change in *Beta-Existence*), the lack of response by the *Physical Universe* not only invalidates the *beam*, but also the *Seeker's* progressive development (*personal certainty*) for *creating energy-beams*. This also keeps many supposed *"magicians"* and *"mystics"* from ever achieving much actual experience in these areas.

The following mental exercise systematically practices *handling energy-beams* without attempting to overcome the *reality-agreements* of this *Physical Universe*. Even so, a *total certainty* or *mastery* is not expected in one sitting or *processing session* with the exercises in this lesson. Many do not have a finite *end-point*, except a *Seeker's* own sense of improvement. Much like we might choose to regularly exercise to gradually build muscle for the physical *Body*, so too are there "spiritual muscles" that require incremental exercise regimens in order for us to regain *spiritual strength* and *ability*.

For this exercise, you *Imagine* (*visualize*) an *invisible energy-beam* hanging suspended in the air in front of you. An *energy-beam* is a sort of *free-standing wave*; it does not, itself, "emit" or "radiate" any *energy*. It also does not need to be "visible" to exist. These kinds of *beams* are used by an *Alpha-Spirit* to generate *motion*, or to "push" and "pull" at things—but that is not what we are doing in this exercise.

The first part of this practice simply requires *getting the sense* that there is an *energy-beam* there. Make it about three feet long, mentally viewed (*visualized*) as suspended in the air, unattached to anything that may be present in the environment.

Once the *idea* that *it is there* can be easily maintained (however "unreal" it may seem at first): repeatedly *alternate* "stretching" the *beam* out to six feet long, then back to three feet. For this exercise to be effective: it is important to not simply *"imagine/create"* the *beam* as suddenly two different sizes. First, you make a *"postulate"* or *"Alpha-Thought"* that it will *be* the size that it should be, and then you mentally "stretch" or "compress" the *beam* as part of the *visualization*.

Depending on how long the *process* is *run*, the *energy-beam* often has a tendency to "snap" or even resist being changed from one of the sizes. Regardless of what phenomenon is encountered, a *Seeker* just keeps working with it until the *energy-beam* is fully under their control—because after all, it *is their creation*. Although this practice seems trite or insignificant, there is more to be gained here than may be apparent at first sight.

HANDLING "ENERGY"

When first *processing* the *advanced-levels* with *Traditional Piloting*, a *Seeker* is asked to *"Look* over the Body..." (with eyes closed, using *"ZU-Vision"*) and see if they notice any *energy-beams* (or *flows*) that are connected to it—or coming *in* to it *from* some direction.

We don't simply occupy a "ring" of *360-degree* perception, but instead, more of a "sphere"—which means, we have at minimum, 129,600 (*360* multiplied by *360*) points of potential "spiritual connectivity." Given the extent or duration of our *spiritual history* (or *Backtrack*): no matter which point or direction is randomly chosen, there is likely to have been a *flow* experienced from that direction.

Of course, this does not mean *every* one of these potential-points (or directions) must be *defragmented*, because an individual is not likely compulsively maintaining *all* of these *flows* in the *present*. It simply means that *any* one of them may be *run*. [The following instructions are adapted for *Solo-Processing*.]

Look over the *Body* (with eyes closed) and see if you get a sense for any *energy-beams* connected to it (from outside). These may only appear as very faint "threads" — or you may only get a vague sense of a *flow*. If you don't *see/sense* anything: simply *Imagine* (*visualize*) an *energy-beam* coming into the body briefly (without putting any *force* behind it) and then "*un-create*" it. Alternate this several times (*creating* and *un-creating*) and see if you get any impressions of *real energy-beams* coming into the *Body*; if not, leave this for a second pass through the course material.

Once you do get a "*real*" impression, it may be on *one* or *many* different *energy-beams* (from different directions). After you start handling them, it is quite possible that more will appear. Each *beam/flow* is either one *you* have *created* or *generated* at some point, or it belongs to someone else. While *running* this *process*, if it is not yours, it will detach and/or disappear; and if it is yours, it will turn a "golden" color when the *energy* is *defragmented* (or "*cleared*," as some *New Age* practitioners say). Also, if it *is* yours, the *beam* may not entirely erase; but it should not seem "vivid" or "active" unless you specifically put your *attention* on it.

The entirety of this exercise consists of *Imagining* (*visualizing*) many *energy-beams* parallel (alongside) to any *beams* you found. It is preferable to use "golden" *beams* for this. If the new parallel *golden-beams* you are *visualizing/creating* start to "snap in" to the "*real*" one, then just keep "feeding" it many more of them until that stops happening — at which point it will *defragment* and/or *detach*. For now, simply clear up any *beams* that are obvious and easy to work with. In more *advanced-level* applications, you can eventually learn to "trace" the *source* of each active *energy-beam*, and find out who *created* it and what it's for (or connected to).

Fragmentation occurs because the *energy-flow* of our *attention* is *unknowingly*, *compulsively*, and *continuously* suspended on "things" — and we call *imprints* "things" because at low-*Awareness* levels, "thoughts" get *charged* with a lot of entangled *energy*, become a "*mass*" and seem quite "solid" as our *reality*. This is what results in "feeling depleted" as a *Being*—because *creation* only gives us things to *have*; it does not, by itself, actually provide us with *energy*.

In many ways, what is considered *energy* is an inversion (or opposite side) of what is considered *mass*. To be able to *handle energy*, one must also be able to handle (and *confront*) *"destruction."* Even on a "mental" or "spiritual" level, *"created energy"* is had by the *destruction* or *"dispersal"* of *"created mass."* Of course, the less an individual is able to handle *dispersal* of their own *creations*, the more "weighed down" they feel—and the less *"free attention-units"* they are able and willing to *Self-Direct*. *"Mass"* tends to restrict *free motion*.

Of course, to *create energy*, the *mass* must be *created* first; but since we have unlimited potential for *creation*, we also have an infinite supply of *energy*. Unlike what we are *implanted* to experience *as* the *Physical Universe*, our spiritual abilities require no esoteric "balancing act" regarding some fixed amount, or "conservation" (as the physicist says) of some *"energy-mass* equation."

Mental Imagery exercises from previous lessons have mainly focused on *"creation"*—and our *processing* has been aimed at handling *"mass"* and *"terminals."* To *handle energy* as a *systematic process*, we have to go about it a little differently. In actual practice, directly *Imagining* (creating) *"energy"*—or *mental images* of *"energy"*—is not an effective approach for *this* present level of *defragmentation* work and personal development.

In order to treat *energy*, we treat *destruction*. An individual must be able to handle (*confront*) both *"creation"* and *"destruction"* in order to regain *spiritual power*. *"Power"* comes as *both* combined—any *fragmentation* or reactive avoidance (*withdrawal*) with one, is to have inabilities with the other. The *power* to *create* requires the *responsibility* and *control* of the same magnitude of *power* to *destroy*. And while we are really speaking of *spiritual abilities*, "thought-forms," and *Alpha* states, if a *Seeker* decides to take a "wider look," they may also see ways that this principle is demonstrated in the *Physical Universe* too.

As another matter of principle: we tend to *create* more than we "need" to be certain of our "*having*" things, or to maintain a certain satisfactory or acceptable condition of "*havingness*" (to be systematically accurate).

As a simple example: a farmer that needs *four* bushels of wheat to survive, plants *ten*. *Two* bushels worth get lost to some pestilence; *three* are sold to maintain other costs; and *one* is given to the *miller* to turn the other *four* into a usable grain-flour. These are not necessarily realistic quantities—the point is: more than *four* are planted. And we maintain similar *considerations* about our *creations*.

Now let us look more directly at how all this applies to *Systemology* and why we cover some of these *advanced* areas in this *Professional Course*. There are a couple basic observations in physical science that ring true for all *Systems*. When *energy* is condensed, it becomes *matter*; when *matter* is dispersed, it becomes *energy*. A *dispersal*—such as an explosion—is really a *series* of energetic "*out-flows*" from a single (central) point. These "facts" haven't changed.

The "*power*" a *Seeker* regains from *systematic defragmentation*, is not earned by simple "erasure" of *fragmentation* and *imprints*. We are actually "*destroying*" these *compulsive continuous creations*—but only *after* giving a *Seeker* the *certainty* that they can *create* any of them again at will. The entangled *energy* is restored "as *energy*" when the *mass* it once was (as an "*imprint*") is *destroyed*. A *Seeker* feels "*relief*" because their *persistent creation* (or *imprint*) is actually "*dispersed*." It doesn't just vanish or disappear; it *explodes*.

Therefore, at *Systemology Level-6*, instructions for "*mental imagery*" or "*handling energy*" involves a lot of *explosions*. And just *Imagining* "explosions" themselves is not effective for *processing*. The *ability-to-confront* "explosions" is critical for many reasons, but for present purposes, let us consider that: if you *Imagine* (*create/visualize*) a "rock" and then "*un-create*" by simply making it "dis-app-ear," there is no *energy gain*. If you, however, *decide* to "*blow it up*," then you've just *created energy* from a formerly *created mass*.

There is no point in *destroying* our *finest creations*—all a *Seeker* needs to do to practice with this is to *Imagine* (*create*) tiny "particles" and make them "explode." Do this many times. Then practice making a *series* of small explosions along a line, as if burning up a piece of string. You can practice with eyes closed, then open. The purpose here is just to *get a sense* of the *energy* it-self.

Once a *Seeker* has accomplished this and feels good about it on a smaller level: *Imagine* (*create*) incrementally larger objects (rocks, buildings, planets, galaxies) and have them explode. Also practice doing this on different sides (directions) of your *viewpoint*—such as "behind you" or "above you," *&tc.*

There are other basic manifestations of *energy* that include our perception of *space*. "*Heat Energy*" is millions of particles moving quickly, colliding and exploding. "*Cold Energy*" is millions of tiny "implosions" that actually slow and stop the motion of other particles. At this level of development: a *Seeker* may "play" with various "*mental exercises*" regarding *energy* and *explosions* freely.

Now let's consider some "*energy sources*" as they exist in *Beta-Existence*. The most basic natural *energy source* that you are familiar with is the local *Sun* for which *this* "*Solar System*" of *planets* is named. One will notice that "THE SUN" is absent from our *tour of planets* in *Lesson-10*. We'll use it now for the next exercise.

A *Seeker* can refer back to earlier instructions for assistance—but the entirety of this exercise is to *get a sense* of being above the *Sun*, looking down at it; and then practicing some *reach* and *withdraw* (mentally connecting and letting go) until you get comfortable with its being there (which is the same as saying that you can comfortably *confront* its *existence*).

Once you are doing fine with being near the *presence* of the *Sun* (which, by the way, is quite different than the *presence* of a *planet*): *Spot* the motion and collision of particles within the *Sun* that is turning *destruction* into *energy*. Get a real sense of the *heat-energy* produced. Then continue your mental *reach* and *withdrawal* with the *Sun* (while maintaining the *sense* or *Awareness* of the activity actually taking place within it).

Now *Imagine* (*create*) a "*copy*" of the *Sun* alongside it. *Spot* any differences between your *copy* and the *Sun*, and adjust your *visualization* by *intention*. When you're satisfied with your *copy*, make another *copy* on the other side of the original *Sun*. Continue to do this until you are comfortable with easily making *copies*.

To advance this further, if a *Seeker* is already well practiced with "*ZU-Vision*" or *spiritual perception*: *Spot* a different "*star*"—preferably one of a different color than the local *Sun*—and once there is a good perception of it, the above exercises are repeated on that "*star.*" Continue to do this with various *stars* until you are comfortable making *copies* of these as "*energy sources.*"

HANDLING "ENERGY-SYSTEMS"

After a *Seeker* is comfortable with the material and exercises in the previous section, the *Systemology* of "*personal energy-systems*" is introduced. In this area, our methods tend to overlap with various *mystical* teachings and practices that one might find in the *New Age*; but again, our treatment of it is *systematic*.

To *get a sense* of "*having energy*" or "*increasing energy*": *Imagine* (*create*) many small "*sun-stars*" (a foot or so in size) in the *space* surrounding you, having them connect to you and "feeding" you *energy*. Continue to do this until you feel comfortable about "*having energy.*" Then, make many even smaller "*sun-stars*" (an inch or so) all throughout the body, *getting a sense* that they are providing *heat-energy.* Do this for all the various parts of the body. [This is the true essence of ancient *"StarFire"* rituals that have since been confused with esoteric hype and blood rituals.]

Not surprisingly, one of the easiest ways of handling personal energy is also the most commonly known. For this, a *Seeker creates* and *in-flows* energy by *Imagining* a "cloud of golden energy" surrounding them, and "flowing it into" the *Body* intentionally using their *"breath."* Although this practice is commonplace, a *Seeker* should recognize that *they*—as an *Alpha-Spirit*—is *creating* this *energy-source*; that it is *not* coming from some obscure *other-determined external* "Divinity" or "Cosmic Source." [In spite of the language used by some *mystics*, if your goal is *Ascension*, there is no *"Cosmic Consciousness"* or *"Universal Mind"* present within *this Beta-Existence* that you are going to want to be "One" with.]

There is an entire spiritual philosophy behind *"breathing."* Most of it simply involves ways of delivering more oxygen to the body; or to regain *conscious control* of the automation associated with *"breathing."* Some of our earlier work in this area may be found in *"The Power of Zu"* volume of the *Systemology Core* (also published as *"Mardukite Zuism: Academy Lectures, Volume 5"*). But that is not essential to what we are dealing with here.

The *"in-flow with breath"* is used in *systematic processing* to "bypass" some of the *energy fields* and *spiritual machinery* that we have set up around us, which typically handles our *energetic* functions on an automatic basis (and which also includes a lot of defense mechanisms to prevent this kind of intentional tampering). [*Flowing-in* large *golden-clouds-of-energy* with *breath* does produce an effect; and it is for this reason that some *magical practitioners* have "stumbled on" sporadic results during varied eclectic *arcane* experiments.]

To do this effectively, a *Seeker* would practice what *mystics* refer to as "pore breathing"—where an individual *visualizes* the surrounding *golden energy* as completely permeating into the *Body* through all of its "pores" with each "breath." As a result, the *breathing* is generally slow and deliberate—but *attention* really doesn't need to be given to any particular "rhythm" or method otherwise.

This type of work leads us into our next area of handling *"personal energy-systems"* directly; and this is one of those subjects that a *Seeker* may have some familiarity with from other traditions and teachings. There is a lot of information available regarding what the *Eastern* practices refer to as the *"chakras"*—and it is by this label they are best known. The only concern in continuing to use the word *"chakra"* here, is if any of a *Seeker's* prior learning in this area inhibits their progress with our presentation of the *Pathway*; but if so, at this level of development, altering the label alone will probably not make much of a difference.

The *enlightened* among us have long understood that individuals in *this Universe* are *"microcosms"*—*"micro-cosmos,"* *Universes* unto themselves—that reflect the same *Systemology* as what we treat as *Universes*.

In previous lessons, we described the *"condensation-of-universes"* that is mirrored in *kabbalistic* representations and other *esoteric* diagrams depicting various inter-dimensional *"levels"* or *"spheres."* By this we mean how *Universes* followed a certain sequence of "collapsing-in" on each other to bring us to *this* version of *Beta-Existence*.

Technically, our *considerations* for these "prior" *Universes* still *exist*; they have simply been "superimposed" upon several times since. The *energy* that went into the *foundations* of these *Universes* is still with us—but it is more *condensed* (therefore more *"solid"*) than what we experienced at *"higher-levels."* But all of the *esoteric* associations—labeling various levels, color-coding them with *"mantras"* (intoned words) and notes of music—have not delivered anyone *out* of *this Universe*. Tables and charts do not substitute *Knowingness*.

The practices suggested are not particularly "difficult"—even reflecting some of what you might find dispensed from a basic newsstand *"How-To"* publication. Still, we do not introduce *"chakras"* officially to a *Seeker* until *after* they've completed *Beta-Defragmentation*. Getting a *"chakra-alignment"* or *"aura-cleanse"* is no stable substitute for the *Pathway*. In a state of *fragmentation*, any beneficial effects will have "peaked" within three days. Without *defragmentation*, the individual would be right back to where they were a week later. Such practices also tend to put the individual entirely at the *effect* end of another practitioner as *cause*.

PERSONAL ENERGY-SYSTEMS

In systematic terms: the *chakras* are *"metaphysical"* (*spiritual*) energy-centers that make up the *"personal energy-system"* that connects *"Bodies"* to *"Universes"*—anchors *"identities"* to *"existences."* Although there are hundreds of these *energy-centers*, we usually use the term *"chakras"* to refer to a specific central *"personal energy-system"* that consists of *seven-plus-one* (*eight*) of them operating as a network.

In our *Systemology*, the *"chakras"* directly align with what we have been working with—a "scale" of interaction (*communication*) between the *Alpha-Spirit* and *Beta-Existence*. This is why, at *Systemology Level-6*, a *Seeker* that is already familiar with our original *Standard Model*, the *Awareness Scale*, the *ZU-Line*, *Spheres of Existence*, and *Levels of Universes*, does not require any additional outside research or cultural-systems to compose (and understand) their own *"Master Chart"* of the *"chakras."* It is the same *seven-plus-one* (*eightfold*) *system* that we have been employing this whole time.

That all being said, we give a *Seeker* some assistance in this area by synthesizing a new *Awareness Scale* for the *"chakras"* that summarizes the aspects of our philosophy (just listed above). Here we pair the data directly with *systematic correspondences* for the *"Standard Model."*

8. CREATION : "Infinity"
7. KNOWINGNESS : "Alpha Spirit"
6. PERCEPTION : "Alpha Thought"
5. COMMUNICATION : "Intention"
4. SYMBOLS : "The Mind System"
3. EMOTION : "Reactive Thought"
2. SEXUALITY : "Organic Sensation"
1. EATING : "Self-Survival"

And here we pair the new *Awareness Scale* with our *"Spheres of Beta-Existence"* data.

8. CREATION : "Infinity"
7. KNOWINGNESS : "Spirit"
6. PERCEPTION : "Cosmos"
5. COMMUNICATION : "Earth/Life"
4. SYMBOLS : "Humanity"
3. EMOTION : "Groups"
2. SEXUALITY : "Home/Family"
1. EATING : "Self"

It is also important to note that this *"personal energy-system"* runs parallel with previous *"Universes"* where we have experienced being *"identified with a body."* These *"chakras"* are still carried with us from our former occupations of *Bodies* that we consider relatively *"more subtle"* in their *solidity*. As long as we have been "using" *Bodies*, we have maintained a *"personal energy-system"* in which to *anchor* them in a *Universe*. The *Systems* for former *Bodies* are "collapsed/condensed" and carried forth by an *Alpha-Spirit* into *this Universe*.

For example: the "survival" emphasis of this *Game-Universe* is based on "preservation of a *Body*" as the most fundamental goal. This follows well with keeping people in conflict for superiority and domination. The underlying principle connecting these two factors is *"To Eat"* (and to a lesser extent, *To Consume*). This is the lowest possible common denominator (or factor) of the currently existing sequence of *Universes*—of which the other *"higher"* ones have since *"collapsed-in"* to. As such, it is directly linked to the *lowest* "root" *chakra*, that is aptly named because it is literally what *roots* or *anchors* us exclusively to *this Beta-Existence*.

We have every reason to believe that prior to the *condensation* of *this Beta-Existence*, the *"personal energy-system"* used in the prior *"Magic Universe"* was linked to a slightly less *"solid"* Body, and *"rooted"* to *that Universe* with what we would now consider the *"second lowest chakra."* This would make sense since the underlying *"game-goals"* of the *Magic Universe* focused much more on hedonistic pleasure: *"To Enjoy."* At our present level of existence, this would be primarily equated to sexual sensation and intimacy.

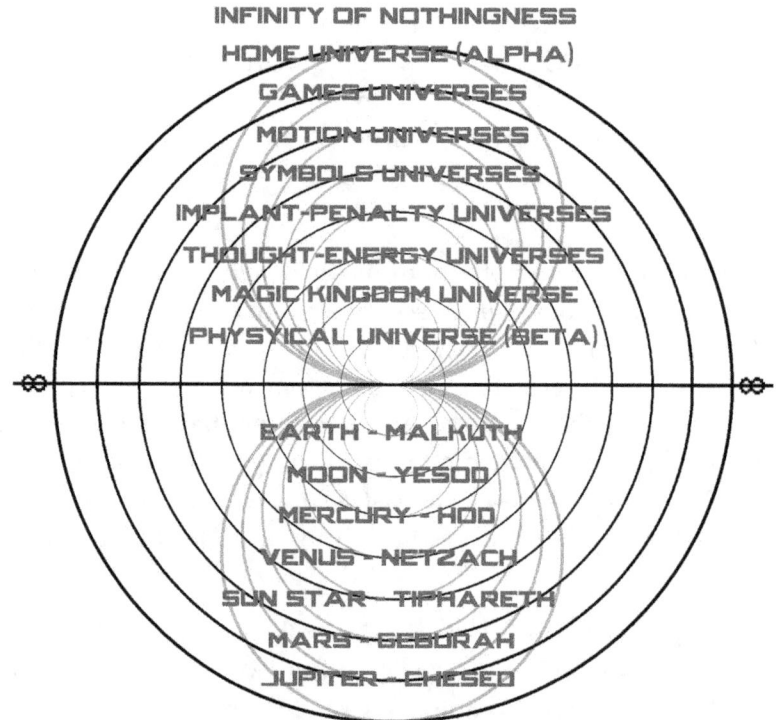

[*Diagram is for illustrative purposes only.*]

In systematically describing *the Physical Universe* and the *Magic Universe* in this way, we have also explained the function (or *"en-*

ergetic domains") of the *lowest* ("1") and *second-lowest* ("2") of the "*chakras.*" It may help to envision the "*root chakra*" as in the vicinity of the feet, "grounding" one's *Body* to the *Earth* (though most interpretations assign it to the region of the "*rectum*"). The *second chakra* is traditionally nearest the genitals or sexual organs (with obvious connections to that department). In regards to *Universes*, these are the two we have the most data for.

The *third chakra* ("3") is located near the stomach. It primarily governs emotions and emotional responses. In relation to the "*spheres,*" it concerns "*group energy*" —or "*synergy*" —which is highly energetic and emotional in content, though not necessarily as intimate as what is handled with the *second chakra*.

The *fourth chakra* ("4") is also known as the "*heart chakra*" because of its location and association with compassion. However, on a systematic level, it essentially governs *energy* associated with "*significances*" and "*importance*" —which at the general *societal* (or *Human*) level of interaction, is represented by "broad symbols" rather than "specific" things.

The *fifth chakra* ("5") concerns our interactions or *communication* with other living beings. It is located near the throat and is often associated with "speech." It is distinctly connected to the "*sphere of livingness*" (or *All Life on Earth*), since living things are in communication with one another much differently than how we are in contact with the inert matter of the *Universe*.

The *sixth chakra* ("6") is usually near the eyes, from which we consider our ability to "*look*" —or else "*perceive*" —that there is a "world" or *Beta-Existence* "out there" in which to interact with. In terms of our *Standard Model*, this is the original "level" of *Awareness* at which an *Alpha-Spirit* first "entered-in" to *Shared/Games Universes*; thus it concerns an ability to *perceive* things apart from our own *creations* and total *Knowingness*. Systematically, this is the point when we first permitted ourselves to "*Not-Know,*" so that we would have something to "*perceive*" or "*know about.*"

The *seventh chakra* ("7") is often referred to as the *"third eye"* (located near the region of the forehead, above the eyes). On all of our models and scales, we typically associate this level with the *Alpha-Spirit* itself, as a *Spiritual Awareness*. But, for practical purposes, this is the same as referring to our original native state of *Total Knowingness*. In terms of *Beta-Existence*, it provides that "intuitive" degree of *Knowingness* of what we would otherwise have to *"look for"* or *"find out about"*—and which often defies *"reason."* [In a *fragmented* state, what usually passes for "human intuition" is really *reactivity* and *imprinting*; not this upper-level of true *Knowingness*.]

Finally, the *highest chakra* ("8") is essentially the *"godhood chakra"*—and we always systematically represent the *"highest"* ideal or aspect as the *eighth* (or with an "8") to demonstrate its *Infinity*. Nothing is *higher* than (or *beyond*) *"Infinity"*; and *"Nothingness"* is *"Infinity"*—but within that state is the "absolute potential" for the *creation* of *"Everythingness"* (as reflected by the actual manifestation existing throughout *lower levels of creation*). At the *highest* level, there is only the *pure Awareness* of "I-AM."

DEFRAGMENTING "ENERGY-SYSTEMS"

The *chakras* are obviously not "physical constructs" that are a part of the "physical anatomy" of the *Body*. The idea that they are "located within" is more figurative or conceptual. They track along, parallel, to the *Body* that we *identify* with as "ours"—but operate at a higher-degree of "spatial" existence. They also correlate with the other *"astral"* layers or *"auric"* body-like *fields*, which we still carry with us from these *higher*/former *Universe* experiences—and which *mystics* and *spiritualists* have been aware of for a long time.

The word *"chakra"* is *Sanskrit* and implies a *"spinning disc"* or *wheel*. Although we tend to conceptualize them as *"balls"* or *"globes,"* that is only how they *appear*. The *"disc"* is simply spinning so fast that it appears quite "*sphere-like*" (like when you "*spin a coin*" on a table-surface). They are also quite "reflective"—much like the *discs* used for music and video medias today. This gives them a quite radiant "metallic" appearance with a slight "sunburst" quality.

To begin systematically working with (and *defragmenting* or *clearing*) chakras as *energy-centers*, it is important for us to get an *actual* sense for where they are "functioning" relative to the *Body*. Although our emphasis is on the *chakras*, we start with a general "*ZU-Vision*" technique for locating something by increasing spiritual perception. Instead of explaining a lot of theory behind the exercise, let's just get right into it.

The technique simply requires *visualizing* "approximations" of a *thing* in various places until it starts to "feel right"—or rather, the *visualization* starts to get "pulled into" its correct shape and location. In brief: this operates on a systematic theory that as your *ability* to *create* a "thing" increases, so does your level of *perception* and *Knowingness* (about that *thing*). We use an "approximation" because this technique is more effective the "closer" you already are to what you're treating. In this case, we're handling *chakras*.

For applying this to *chakras*: *Imagine* (*visualize/create*) "*spinning discs*" of *energy* around the *Body*. For this, we *don't* start by making them *in* the *Body*, so that we are not in *conflict* with *actual energies*. So we start with a bit of "testing" to try and *perceive* what *is* there rather than fight with it or force anything with our *visualizations*. You just keep placing these "*spinning discs*" of *energy* on all sides around you, above and below. As you keep doing this, you may *get a sense* that they should look or be a certain way; in which case, you go ahead and start making them that way.

After a *Seeker* has done this for a while—given that the "*intention*" and "*attention*" has been on *locating* and *perceiving* the *chakras*—the created "*spinning discs*" should develop a tendency to get "pulled in" to the *Body* (or possibly "snapping in") at their appropriate locations. Rather than change what you're doing, just keep "feeding" the *Body* these *"spinning discs"* by continuing to put them around the *Body* and allowing them to naturally get "drawn in." Eventually, *perception* on the *actual chakras* should gradually increase. A *Seeker* applies *"ZU-Vision"* primarily for this, with eyes closed, adjusting their *spirit vision* (by *intention*) to the "frequency-band" or *level* of the *chakras* while *looking* at the *Body*. This exercise is practiced until a *Seeker* is satisfied with their ability to *perceive* the *chakras*.

Now there is the matter of *defragmenting/clearing* the *chakras*, which itself requires some explanation. The *chakras* are essentially parts of a *"personal energy-system,"* but it is not a "closed-system." These *spinning discs* or *wheels* are just one part of the *"spiritual machinery"* we maintain a *creation* of *unknowingly*. While we experience *Beta-Existence* at its "normative" *Human* range of sensory-*Awareness*, the *chakras* also automatically "cycle" or "process" the actual *spiritual energy-flows* that we are interacting with. Of course, each *chakra* tends to filter its own specific "area/level" or "domain" as we've described by numbering them.

Spiritual "health"—for example, the "condition" of the *chakras*—is closely integrated with the general state of *fragmentation* or *Self-Honesty* that an individual maintains or experiences. This means that a lot of what a person hopes to "fix" with *chakra-clearings* or *aura-cleanses* is really handled with *defragmentation processing* in our tradition or philosophy. In our opinion, our methods provide more stable long-lasting results by correcting how an individual is *operating*, rather than spending all one's *attention* on constantly *cleaning up*.

The *chakras* each have the basic characteristics (or domains) as described in the previous section. These are simply areas of potential *fragmentation*. Some *mystics* describe various "overactive" and "underactive" qualities—but it really all adds up to *fragmentation* connected to each of those domains. For example: issues with physical illness pertain to the *first chakra*; relationship problems or perversions primarily denote *fragmentation* of the *second chakra*; and so on. Therefore we can also list "opposite" characteristics of an *inverted-Awareness Scale*, each demonstrating the extreme *antithesis* of a specific domain.

(+8) Creation; (−8) Destruction
(+7) Knowingness; (−7) Mystery
(+6) Perception; (−6) Delusion
(+5) Communication; (−5) Isolation
(+4) Symbols; (−4) Barriers/Enslavement
(+3) Emotion; (−3) Misemotion/Hate
(+2) Sexuality; (−2) Perversion
(+1) Eating; (−1) Physical Illness

The present author spent *thirty years* pursuing esoteric mysteries and arcane teachings. While the actual existence of these *energy-centers* is well observed throughout many traditions and many labels, it has become clear that there is no consistency between various interpretations of the "colors" or "geometry" of the *chakras*. As far as can be determined, their structure is likely to be *perceived* differently by different practitioners. In our tradition, we apply methods that were developed for *20th Century American New Thought*; and these may differ from what you may have learned from other "*New Age*" sources.

For example: we treat all of the *chakras* as a "*golden*" color, but with a faint prismatic rainbow effect as the *disc spins*—like what you see on a *compact-disc* or *DVD* surface when you angle it in the light a certain way. This also makes it easier to treat them all with the same style of technique that we've already been using.

The type of *fragmentation* that "*spiritual energy clearing practitioners*" describe will usually appear like black spots or dark discolorations. This is the "*energy-mass*" of *fragmentation* that is restricting an otherwise *clear channel* of the appropriate energy for that *chakra*. We handle this in *Systemology Level-6* to mainly "clean up" residual effects, because we have been indirectly *defragmenting* the *chakra* "domains" all along the *Pathway*.

While you can technically apply some *attention* toward a "dark area" and get it to turn white by *intending* or *creating* a small stream of neutral or golden energy into it, this does not solve what *caused* the misalignment in the first place. Introducing these methods too early might just make an individual *more comfortable* with their *fragmentation* rather than handle it. *Defragmentation* only comes about from actually *confronting* something with *Awareness*; certainly not by *withdrawing* and pretending things aren't there.

We will end this lesson with a *systematic processing* technique for *defragmenting* the *chakras*, based on the descriptions provided to a *Seeker* previously in this lesson. This should also only be handled *after* a *Seeker* is comfortable with the other exercises given previously in this lesson.

A. *"Close your eyes; alternately spot three points in the chakra, and three points in the room."*

B1. *"Spot things in the Universe that the chakra might be connected to."*

B2. *"For each thing: alternately spot the chakra and the connection to it."*

C1. *"Look for areas causing 'dark spots', or turbulence in the energy-flow, and spot what they are connected to."*

C2. [*Handle the fragmented area directly with one of the processes in the Professional Course; or simply handle it as a 'protest' as given in the next steps. See also: Lesson-4.*]

D1. *"Spot what is being 'protested' in connection with the area."*

D2. *"Alternate 'protesting' and 'admiring' it, until the energy begins to flow freely again."*

The general guidelines (about *chakras*) provided within this lesson are the most common workable factors that a *Seeker* may incorporate, and have actual certainty on, at *Systemology Level-6*. There is obviously more to each of the areas our *Professional Course* covers, but we have focused on the most critical points necessary to compose a complete and workable *paradigm* or *applied philosophy*.

LESSON FOURTEEN
"SPIRITUAL MACHINERY"

MACHINERY AND CIRCUITRY

In the previous lesson, we introduced the subject of *"energy"* and *"energetic systems."* Now, we advance this further by considering how *"matter"* (*forms*) and *"energy"* are combined to form *"machines"* and *"bodies."* By *"spiritual machinery,"* we mean mechanisms that are *"actual constructs"* carried by an *Alpha-Spirit* along their experience of *spiritual existence*. And, while they may not be *"physically apparent"* or perceived directly within the sensory range of *this Physical Universe*, they very much *exist* and *affect* us. A perfect example of this, is what is often referred to as a *"Mind."*

There are two main types of *spiritual machinery*. There is *light machinery* that is simply *created* as itself, as a *form* with a particular *function*; and there is *heavy machinery*, which is more "organic" in nature, because it is actually built out of *spiritual beings* (or else requires *an entity* be *entrapped within* it in order to function). For now, we will focus on *light machinery* as an introduction (leaving the subject of *entities* and *spiritual-identity fragments* for a more *advanced level* of personal development that follows *after* the *Professional Course*).

In basic terms, when we refer to *machinery*, we mean a *form* that is distinguished by its *programmed functions*. It *does* something; but it only *does* what it's *programmed to do*. In the case of our own *"Mind-System,"* we mean a complex network of *energy-systems* that give the *"Mind"* its *form* and *function*. It is a vast array of *"circuitry"* that operates *automatically* on a *"push-button"* or *"stimulus-response"* basis.

The *"Mind"* is a *creation* of the *Alpha-Spirit* that continues to *persist* and is left to *create* for it *automatically*. The longer it *persists*, the more *"energetic-mass"* it seems to contain as a *form*, meaning the greater of a *"weight"* it becomes for our own continuing *spiritual existence*. *Traditional Piloting* requires communication with the *Alpha-Spirit*—the *actual individual*—and not some *social-machinery circuit* that when you ask *"how are you,"* it is *programmed* to automatically respond *"fine,"* even when the person is not. That is not *true communication*. This is why *actual presence in-session* is required for *systematic processing* to be effective.

Upper-Level Systemology is sometimes difficult for a *Seeker* to approach and understand without having already increased their *spiritual perception* (or *"ZU-Vision"*) with *Beta-Defragmentation*. Effectively working directly with these relatively "higher" areas requires more *actual intuition—Actualized Awareness*.

Starting with *Systemology Level-5* (*Lesson-11*), we began dealing more with *"things"* (*terminals* and *forms*) and *"incidents"* that are not *directly* encountered (or typically *visible*) from *within* the normative "confines" or "limitations" of the *Human Condition*. Although they continue to affect our experience of *Beta-Existence*, they *perturb* and *impinge upon* our *reality* as an *"unseen source"* of various *creations* and *manifestations*.

In more *advanced* work, *"Biofeedback Devices"* may be used to detect when *fragmented charge* in certain *"unseen areas"* is present. So, what we are speaking of is not entirely "undetectable," just not always *knowingly* sensed, especially when it is *"active"* or *"restimulated."* [Our goal with the *Professional Course* is to take a *Seeker* as far as they can go *intuitively*, before introducing *advanced* applications of *GSR-Biofeedback, &tc.*]

There is nothing inherently wrong, or even *fragmenting*, about an *Alpha-Spirit knowingly creating spiritual machinery*. This is completely within the native abilities of *Alpha-Thought*. An individual can certainly make a *series of postulates* and *program* an *"external* form" to *automatically create* an *effect* when some *programmed* factor "triggers" or "activates" it—meaning, when something *"pushes the button"* so to speak. This is actually, in part, *how* we even started *creating Universes*; and then later, by *creating machinery* that could then *create* and automatically and invisibly *manage Universes*.

Much like any other *creations* and *"postulates"* an *Alpha-Spirit* makes: our troubles only come when we set our *creations* to *persist* (run forever), forget about them, and then lose all *responsibility* and *control* of them. This is essentially what *produces all* of our *fragmentation*. When an *Alpha-Spirit* fully regains *responsibility* and *control* of its own *creations*, it will also regain more of its own native *ability-to-create knowingly* once again.

Advanced philosophies aside, let's get busy doing some practical exercises for this lesson.

SOME BASIC EXERCISES

A *Seeker* develops and applies their *visualization* and *mental imagery* skills all along the *Pathway*. At this *processing-level*, more and more of the exercises require that ability. A *Seeker* must be able to actually *get the sense* that they are *doing* something, or *creating* something, with their *intentions*—even in the absence of having a directly observable and solid manifestation.

We have practiced many exercises with the *Body's Eyes* closed, but here we will develop and apply the same principles with *eyes open*. One should work with this until there is a certainty on being able to clearly "superimpose" one's own *mental images* or *creations* "over" the *Universe* that is seen with the *Body's Eyes*. In previous lessons, we mainly use solid simple objects. Now we want to add some complexity to it. [There is also no expectancy for a "physically solid object" to suddenly materialize for others to see. That is not the intention; and what's more, there is no "postulate" made in this exercise for such to even occur.]

A. *"Look around (the room); Spot an object you like, which has a bit of complexity to it."*

B. *"Imagine (create/visualize) a duplicate-copy of the object right next to it."*

C. *"Alternate your attention between the two objects, one to the other; Spot any differences between the two and adjust your copy to more closely match the real one."*

Once a *Seeker* is fully confident doing this, the same exercise is practiced with *eyes closed*. However, this time, use *"ZU-Vision"* (or *"Imagination"* if necessary) to *"Look"* around and *"Spot"* an object that is *outside* the room, preferably *outside* the building in which you're holding your *session*. If you're working outdoors already, put your *attention* on *"Looking"* at a *location* that is not close-by. Whatever it is you find to work with, make a *duplicate-copy* next to it and keep adjusting it until there are no noticeable differences.

Depending on a *Seeker's* previous development, it may require one or more *sessions* to be fully certain of ability with these two parts (above). For the final step, alternate between the two parts using different objects this time. With *eyes open*, pick an object in the room to *copy*; then with *eyes closed*, pick a large complex object that is *external* to the *room* (requiring a *viewpoint* that is *"remote"* from the *Body* in order to *"see"* it) and *copy* that. Then throw away or destroy all your copies and end the *process*.

The remainder of our basic techniques in this area (given in the next section) were first introduced to *advanced Seekers* many years ago in our original *"Wizard Training Regimen"* for *Systemology*.

FACSIMILE-COPIES: MACHINERY

• Select an object that has the basic mechanical function: "*to produce a flow.*" [This may be initially practiced with a "sink" or "water-spigot" in the absence of a reality on *electricity* or understanding of *motion.*]

–With eyes open: *Look* at the mechanical object in the "OFF" condition; *Imagine* an identical duplicate beside it. *Look* between the two alternately; *Spot* any differences and adjust your *created-duplicate* to match the original. Continue until you're satisfied with the *certainty* of the *duplication.*

–Turn the mechanical object "ON"; *Look* at it in this condition, *noticing* the motion, and *getting a sense* of the *energy-flow* driving it. Adjust your *imagined duplicate-copy* to match this in every way.

–Then alternate these steps, *duplicating* the functions "OFF" and "ON."

• To advance this further: apply the previous steps, with eyes closed, using an object not present.

• Select a mechanical object that has a basic "motor" function. [An "electric fan" is best for practice until a *Seeker* has a greater reality on *generators* and *engines.*]

–Apply the basic steps (previously) for *imagining duplicate machinery*; this time *noticing* the internal mechanics: a *circuit* or *energy-flow* that drives or propels the *spinning-motion* of the *blades* that is *started* and *stopped* (*controlled*) with a *switch.* As it "runs" (is "ON") *get a sense* of the internal mechanics, matching this *energy* and *motion* in your *duplicate-copy.*

• For additional practice with the above steps: use more complex machinery; use machinery not present; use electronic devices. [A basic physical understanding of the way things work is quite beneficial. We recommend David Macaulay's "*The Way Things Work*," which is illustrated and easy-to-understand.]

• With eyes closed: *Imagine Being* a *"motor-vehicle"*; *create* the machine, the internal mechanics, and *get the sense* of *"identifying with it"* as a *body*. Establish a *"point-of-view"* (POV/*viewpoint*) within the *vehicle*, while maintaining the *creation* and a *sense of* the *energy* and mechanical *motion* taking place inside of it.

–As an additional step: move your *point-of-view* through each of the *vehicle*'s mechanical systems, while maintaining a *sense of* it running: steering, brakes, the engine, transmission, &tc., to the extent of your reality on these systems. Even without previous knowledge, see if you can *get a sense* for how it operates *from the inside.*

In general, an individual will have many types of *"mental machinery"* active and running on automatic. In the following exercises, we simply want to *get a sense* for what some of that would be. After all, we either *created* it and forgot about it, or else we were *implanted* with it; but regardless its *ours* now—and it continues to *persist unknowingly* using our own *spiritual energy* as its *source.* This is not likely to be eliminated fully with this *Professional Course*, since its handling requires *advanced processing* used for *Alpha-Defragmentation.*

One setback to working with *mental imagery* and *spiritual energy* is the lack of "feedback" one gets from the *Physical Universe* to validate any gains in *spiritual perception* and *ability.* The following exercise "pretends" that we are *creating* and then simultaneously an *"uncreation"* occurs automatically, due to one of our *mental machines.* It's perfectly fine to treat this as a "theoretical" condition for this exercise without adopting it as a belief (prematurely to *knowing* it).

• *Imagine* (*create/visualize*) a large object, such as a piece of furniture, out in front of you. When you are satisfied with its form; *unmake* (*uncreate*) it. This doesn't mean keeping a *sense of it* there and just making it "invisible"; you apply the same level of *intention* you used to *make* it as *unmake* it. We practice by doing this *knowingly* and *intentionally.*

• *Imagine* that you have an *"unmaking machine"* above-and-behind one of your shoulders; that whenever you *make* the piece of furniture, the *"unmaking machine"* is making it "vanish" (*cease* to *exist*). Repeat this a few times with the same object. Then use different objects. Then do some more, but shifting the location of the *"unmaking machine."*

–Repeat the above step, but using a *"blackness machine"* (which overlays your *creation* with a *"cloud of blackness"*).

–As before, but using an *"invisibility machine"* (turning your *creation* "invisible" when you make it).

• [Using the data above] *Imagine* (*create*) the furniture; use the *"invisibility machine"*; but this time, also *pretend* to *"Be"* a *"looking machine"* inside the *Body's Head*, which can *look through* the *invisibility* (and *see* the furniture anyway) and produce a *picture* or *mental image*.

–A *Seeker* practices the above step until they can do all parts as a single action: *knowingly creating* the object; *making it invisible* with one machine; *looking at it* as another machine; and ultimately *seeing* the *image picture* that the second machine produces. Then practice some more with different objects.

–Practice as before: using the *"blackness machine"* with the *"looking machine."*

• In a public place: *Look* around; *Spot* a "wall" or large object. *Spot* people; *Imagine* them having a *"reality machine"* that projects the *reality* of the selected object/wall.

–As above; then *Imagine* that the *"reality machines"* also *"copy"* the *reality* from one person to the next (so that they all see the same *reality*).

–When a *Seeker* is comfortable with these steps, they can continue the exercise with different objects.

–Finally, practice on the "entire physical reality" that is viewable: *Imagining* that each person you *Spot* has a *"reality machine"* that projects the *reality*; and having each *machine copy* sent to the other *"reality machines"* (so that everyone experiences the same *reality*).

BODIES: PHYSICAL & OTHERWISE

Systematically speaking, a *body* is a *symbol* that an *Alpha-Spirit* uses as their *"playing piece"* in a *game*. It is a *"terminal"* existing at a particular level of solidity or reality that is used to relay communication, perception and action. We often use the term *"genetic-vehicle"* to denote the organic *"Human Body,"* because obviously an *Alpha-Spirit* has used *other* forms as *"game tokens"* previously on their *Spiritual Timeline* (*Backtrack*).

Earlier in our existence, we were able to simply *create* and *uncreate* *"bodies"* at *will*—or an *Alpha-Spirit* might leave a *"body"* in a known position so that others could contact it. In fact, even on Earth, there are reports from the ancient world about *"statues"* that *"gods"* once used to communicate from a distance. But early on the *Backtrack*, we did not *confuse* our actual *identity* as *"Self"* with these *token pieces*; and we had no difficulties or limitations in *creating* or *projecting* "bodily-forms" when we needed to.

As *games* became more interesting and more complex, we required more complex and restrictive *playing tokens* in order to participate. What began even as "simple objects" quickly evolved into more elaborate, dynamic and ever-changing *lifeforms* as *bodies*. At first, we still did not make the mistake of *confusing "Self"* with the *playing token* we were using. So, what happened?

• The next time you *play* a *board-game* (or *video-game*), *consider* how far "downscale" your *Awareness* would have to sink in order to forget that you were a *player* in a *game*, and actually believe that *you are* the *playing piece* that is being moved around, and even forget that what is happening is a *game*. What type of scenarios or sequence of events might bring this about?

We have all come down to this *Beta-Existence* through a long series of *Universes*; and each one involving the usage of specific *spiritual energy-systems* and *spiritual machinery*, each leading to more *"cond-*

ensed" (or relatively more *solid*) layers of "*subtle*," "*etheric*" and "*astral*" bodies—and with each, adding a lower and lower *energy-center* to our *personal energy-system* (referred to commonly as "*chakras*" in the previous lesson).

Prior to *this Physical Universe*, the lowest denominator of *Beta-Existence* was the "*Magic Universe*." When *that Universe* was considered "*Beta*," everything "above" it was *Alpha*. Once *this Physical Universe* came into being, then the *Magic Universe* became a part of *Alpha*. Everything shifts a level with each *Universe*.

In the *Magic Universe*, our lowest "*chakra*," was near the level of "*genitals*" (what is now the second lowest). The lowest one that we now recognize today for the *Human Condition* (the "*rectum*") was likely added during the transition from the *Magic Universe* in order to *anchor* the more *condensed* (*solid*) conception of a "*localized body*" that is used here in *this Physical Universe* now. The entire "*chakra*" system is solely designed to "*anchor*" the *viewpoint* of an *Alpha-Spirit* (as *Self*) to a "*local*" *body* that is in turn *anchored* to its own level of *Beta-Existence*.

• The next time you have access to some *game pieces*, *figurines*, and *dolls*, give yourself a chance to actually *play* with them. Just be *imaginative*; get the *pieces* interacting with one another; essentially do what a child might do with them.

– Select the *figurine* or *doll* you like best and *get a sense* of being *located inside* the "*doll*" (or whatever)—*seeing* from its *viewpoint*, *sensing* the *emotions*, *&tc*. Then, return to your original *viewpoint* as *yourself*; outside ("*exterior*") to the *game*, and "*senior to*" (or "*above*") it. Alternate: "*Being*" the *game piece/doll* and "*Being*" *yourself*.

This type of *play* is most effective as a practice if the *Seeker* is able to continue altering their *viewpoints* while manipulating the motion and interaction of the *pieces/figurines/dolls*. You physically continue to move the "*pieces*" around for the *game*, while alternating between your own *viewpoint* and that of the one selected *piece*.

It may take some practice (several *sessions*) to comfortably shift *viewpoints* until you can reach a point of simultaneously being the *piece* and the *operator*.

One of a *Seeker*'s goals for *Systemology Level-6* is to take this type of practice to an *end-point realization* where you can *perceive* or *recognize* this phenomenon taking place in the *Human Condition*; and to be able to maintain a *viewpoint* as an *Alpha-Spirit* (as *"senior"* or *"exterior"* to *Beta-Existence*), while simultaneously operating the *viewpoint* of the *"Human Body"* as your *"playing piece"* in *the Physical Universe Game*. It will likely require a second pass through the material of the *Professional Course* to get full *certainty/reality* on this.

• With eyes open; outside; in public: *Spot* a person that will be standing or sitting a while (such as at a bus-stop) so that you can practice the exercise without them leaving. *Imagine* (*create*) an identical *facsimile-copy* beside them. As with our earlier similar exercises: *look* at each alternately, *noticing* any differences and adjusting your *duplicate* to match. Then, *Spot* a different person and repeat.

–Additionally: if the person does leave your view during the practice, simply select another. If the person changes their position or location in the area, simply adjust your *duplicate* to match the motion.

–To advance this further: give particular *attention* to *copying* the internal parts of the *Human Body* (bones, organs, muscle, skin) and *get a sense* of the *organic-machinery* functioning inside (as with our previous exercises on *machinery* and *vehicles*).

–Finally: practice these steps to *create a facsimile-copy* of a person in motion (*copying* the motion in your *duplicate*). *Get a sense* of how the internal *organic-machinery* is operating various "motor-centers" and "functions" during the motion. Practice this repeatedly on several different moving persons.

An additional practice would include applying these same exercises to *"animal bodies"* (for example, on a visit to a zoo). If a *Seeker* decides to do this, begin by making a "connection" with the *animals*. *Spot* them directly and silently acknowledge them for being there. *Get a sense* of them acknowledging you in return. It is best to practice with a few different *animals* to avoid *fixation imprinting*. Various potential *realizations* can also be gained by *Imagining* "*Being*" various *animal* forms (as a *secondary viewpoint* exercise).

HANDLING "BODIES"

Another checkpoint of *Systemology Level-6* is for a *Seeker* to have some actual *spiritual perception* ("*ZU-Vision*") and an ability to *look* at their *Body* from a nearby *viewpoint* (that is *remote from*, or *exterior to*, it)—even if only vaguely. If this is not the case, then the exercises in this section are better left for your second pass through the course. Most of these exercises may be practiced with *eyes closed*.

NOTE: These exercises operate starting from a *viewpoint* that is *behind* the *Body, looking downward* at it.

• *Look* at the *Body* and *Imagine* (*create*) a *facsimile-copy* alongside it. This should be an identical *duplicate*; not a mirror-image opposite it. As other similar exercises: *notice* any difference and adjust your *copy*. Repeat this until you are comfortable with it.

–Then, as before: but reorient your *viewpoint* so that you are *looking downward* at the *Body* from a different position; do this for all relevant directions (from each side, from the front, from above and below).

–Then, repeat the above step, but changing the position of the *Body*: if you were sitting up, then practice it lying down, &tc.

• *Get a sense* that you are *behind* the *Body*, extending your *mental reach into* the *Body's Brain* using some kind of *energy-beam* in order to *control* it.

–Do this step (above) with *eyes open*; but still maintaining some sense of also *being behind* (or even to the *side* of) the *Body*, and *looking* at it as you *operate* it.

–Make a simple slow motion with your hand; *get a sense* for the *energy-flow* and *"nerve-impulses"* between the brain and the hand, causing it to move. Repeat several times, *noticing* how the *energy-flow* works. Repeat this step with your other hand a few times; then switch hands and do it again.

–Once you are comfortable with the previous step: start with the first hand, and as you perform the step, put your *attention* on *noticing* if you *get a sense* for any *"barriers"* or *"obstructions"* to a smooth *energy-flow* along the *channel* between the *brain* and *hand*. **If you do: *Spot* two *points* just to either side of the *obstruction*, and using *intention*, gently *flow* some energy back and forth until it *"dissolves"* (or you don't *sense* it there anymore). [Anything you *"see"* is based on your *level* or *range* of acceptable *perception* and not likely to be exactly how these *energies* and *barriers* would "appear" if they were actually visible to normative senses.]

–Move the body some more in this wise and see if there are other *"barriers"* and *"obstructions"*; applying the technique (above) for "if you do."

An *Alpha-Spirit* has many *structural programs, energy-systems* and *mental mechanisms* hooked into a *genetic-vehicle (Body)* in *Beta-Existence.* We refer to it all as *"spiritual machinery"* because, while these things do not have any real material substance in *the Physical Universe*, they do have a material-like structure at an *Alpha* or *"spiritual" ("metaphysical") level of Existence.*

On the *Standard Model* and *systemological charts*, we codify and systematize *seven-plus-one "levels"* between our experience of this *Beta-Existence* and our own native state as *Alpha-Spirit*. The sequence of *condensation* found in *"levels"* of *Universes* is actually mirrored in the *condensation* and *solidification* of *structures* and *machinery* that is actively at play.

Mystical models of *kabbalah* and *Star-Gates*, esoteric doctrines on *chakras* and *subtle bodies*, and various spiritual *books of the dead*, have all pointed toward a higher unified understanding of existence. It is unfortunate that the *"thinkingness"* of the *Human Mind* has prevented that true "incommunicable" and "unspeakable" unified understanding from unfolding for the initiates of *magical orders* and *mystical groups.*

Too often, the *mystic* and *magician* will become distracted by the pomp of classification and labeling. Unnecessary complexity, cultural flavors, and improper communication, all hinder the true personal development a *Seeker might* have gained from pursuing other routes. Many of the other possible avenues provided *some* progress; but when they eventually looped back to our original *Pathway,* we found that the detour was not a short-cut, but actually took longer to trek, and did not get us as far toward the goal after all. [This assumes a *Seeker* doesn't get lost altogether, early on, when chasing down these other paths.]

The *"Astral Body"* and the *"chakras"* originated with the *Magic Universe*—the next *level* "up" or "higher." It is these subjects that most *New Age* practices employ as an entry-point to "spiritual development"—without any *consideration* for the *steps* we've taken on our *Pathway* leading up to this point.

As a result, the *mystic* attempts to use their imperfect knowledge of these *energy-systems* as the total basis of a methodology toward "Ascension." The *magician* often goes one step further in assigning unnecessary *god-names, grimoire hierarchies,* and *mythological pantheons* to the process—further removing *Self* from *Cause,* and yet simultaneously thinking that all the superfluous data will somehow give them greater *control* of it all.

While we introduced our basic *systemology* of the *"chakras"* in the previous lesson, there is another *arcane energy-system* that originates from an ever *higher level* than the *Magic Universe*—and is therefore less *"condensed"* (*solid*) and easier to work with. Because this system consists of only *"golden orbs"*—and not a rainbow-array of fancy geometry (like the *chakras*)—it was likely easy to overlook among the many tears and surviving remnants of the most ancient esoteric lore on the planet.

Although the *chakra-system* is the most readily recognizable from contemporary metaphysical studies, we mentioned (in the previous lesson) that there were many other *"energy-centers"* or *"anchor-points"* in the body. While the *chakra-system* seems to mirror the *"ZU-Line"* and parallels the *central nervous system* of the *Human Body*, there are other *energy-centers* (*"golden orbs"*) that are positioned, for example, near the joints and organs—and even a few (usually *eight*) key locations a couple feet *outside of* the *Body*.

To start with: the *three largest "golden-orbs"* are *in* the *"head."* They are "anchors" because they "track" with a *Body*. They are not materially *inside* the *physical head* in *this Physical Universe*. They *exist* at the *level* that they originated; but they continue to "track" along with the *Alpha-Spirit* in the relative region of whatever it is *identifying* with for a *head*.

Since the *"Astral Body"* (from the *Magic Universe*) is formed quite closely to the "humanoid" form, it would seem that having a form with a *"head"* was preferable for a few past *Universes*. Prior to this, an *Alpha-Spirit* may have chosen just *"a golden orb"* itself, or even *"a nebulous cloud,"* as its *"contactable body"* (as would have been appropriate for much earlier *Universes*).

The structure of the *"golden-orb"* system is likely a couple *levels* (*Universes*) *"above"* the *Magic Universe* and the *"chakra"* system. We lightly introduce a *Seeker* to both *systems* only after earlier *processing levels*. Spending too much time (emphasis) on one or the other (or any) existing *structural energy-system* tends to increase an individual's *"reality-agreements"* with *that system*. This is why many *"New Agers"* get out of one *"box"* and find themselves in just a slightly larger more interesting *"box"* (but still in a *box*).

Much like how we handled the *chakras* (in the previous lesson), we will handle the *golden-orbs energy-system* by "approximation." The three *golden-orbs* in the *head* are the easiest to start with. They fit inside the *head* on a single plane next to each other, like *billiard balls* racked on a *pool table*. Although they are *"anchor points,"* we refer to them as *"orbs"* because (at least in the *head*) they *"appear"* much larger than a single *"point."* The ones at the critical joints (where the *Body* bends) are a bit smaller, as are the eight that are a couple feet outside the *Body*—and then there are also thousands of tiny ones throughout the whole system.

As with the *chakra-system*, handling the *levels* of our *"personal energy-system"* can certainly make the *Body* *"feel better"* (and there are some other interesting effects that a *Seeker* may discover), but it is applied only after (or during later parts of) *Beta-Defragmentation procedures*, for additional "clean up" and "polishing."

A *Seeker* that worked with our *"chakra"* technique (in the previous lesson) will find this exercise easily workable. Again, we *Imagine* (*create*) lots of *"golden-orbs"* around area of the *Body* (such as the *head*). We don't put them inside the *Body* or apply any force to get them into a particular position. With *attention* on these *"anchor points"* specifically, as you place your *creations* in the vicinity, you should *get a sense* for where the "real ones" are, even if only vaguely. The idea is to intend for the "real ones" to *pull in* the copies you are *creating*. [For this exercise, *pushing in* with *force* may result in a *headache*.]

The *"golden-orb system"* originates from earlier *Universes* on the *Backtrack* than even the *chakra-system*. Although it is designed to *persist*, its structure may seem even more decayed, discolored and misaligned than what is experienced with the *chakras*. But it may not. Some of the "repair" naturally occurs during *defragmentation processing*, but that alone doesn't always provide a *Seeker* the knowledge that these *systems* exist.

Similar to how we handled the *chakras*: as the *"anchors"* in the *head* become visible, they may not be very golden; so you can gently *flow energy into* them and get them back to their proper vibrant shining color. As with *"defragmenting energy-flows"* in other *syst-*

ems, you *free up* any "blockages" or *clear* any "dark spots" by *intending/creating* more *clear energy-flow* into the *system*. It's simply a matter of *controlling* the *system* — but this, of course, does not mean *"forcing"* anything.

The same principles apply if you *get a sense* that these *golden-orbs* are "misaligned" (*out-of-position*) — and they most likely will be. Lightly add more *energy* and "encourage" them to shift back into their correct positions (without *pushing* or *applying force*). We handle these *systems* in this wise because it is the *systematic* approach. Since we cannot be absolutely certain of our *perception* of any of these *"higher-levels,"* we may not know exactly where they are or where they should be. We handle it with *attention-energy*; just as we might apply *attention-effort* to turn "ON," "OFF" or *"change"* the functions of some other *mechanical system*.

FURTHER TOWARD INFINITY

We often refer to *reactivity* as a *mechanism*, which is to say *machinery*. We refer to *response-mechanisms*, *defense-mechanisms*, and all kinds of *automatic reactive-mechanisms*. So, from the beginning, we even introduced our *systemology* of the *"Mind-System"* — or the nature of the *"Mind"* — as *spiritual machinery*.

Our *systematic processing* is actually quite easy to use, and is based on very few critical fundamental principles — but one of the most basic ones is: *get a Seeker* to *knowingly do* (or *create*) what they are already *doing (creating) unknowingly* as a means of putting it under greater *Self-Control* or *Self-Determinism*.

The supposed psychological authorities and ignorant critics of our techniques will argue that telling a person to do what they are already doing is a way to lay in manipulative control and brain-washing. And while they are *half-right*, these individuals are probably in most need of getting *processed* in this lifetime.

The truth is that, in *Traditional Piloting* we are *not* having a *Pilot* repeatedly command a *Seeker* to "sit in a chair" when the *Seeker* is already *sitting in a chair*. That would just be stupid. The technique in question is to have a *Seeker* knowingly as themselves (as *Self*) *get the sense* of *making* their *Body* "*sit in the chair*" as a preliminary step to greater *Self-Direction* and *Self-Determination*.

Systematic Processing requires that a *Seeker* put greater *attention* on making *deliberate actions* that are otherwise handled on *automation* and *stimulus-response*. This would be a "step up" for the kind of person that almost obliviously stumbles in and just plops down in a chair. Some beginning *Seekers* can barely *recall* much about how they got to our *office* before they are suddenly *present in-session*. It is obvious that many *fragmented* individuals are essentially *operating on auto-pilot*, to use an appropriate bit of slang.

That all being said, we will continue here with the next exercise without much additional explanation. Some of the underlying "*systemology*" here should now be more apparent to continuing *Seekers/students*.

• *Go around* the *room* and physically touch things; but as you touch each one, *Imagine* the *walls* yelling "*Mustn't Touch!*" at you; and then you let go (*withdraw*) very quickly.

–Eyes closed; repeat as above without moving around the room: this time "*mentally*" *Spotting* a thing; *reaching out* and *connecting* to it; having the *walls* yell "*Mustn't Touch!*"; and you rapidly *withdrawing*.

–Repeat each of the above versions (*physical* and *mental*): this time, when the *walls* yell "*Mustn't Touch*," you ignore it (continue to touch the object) and decide when you want to let go (and then do so).

Often times when a *Seeker* has admitted to "not seeing anything" when they close their eyes (for example, in terms of "*ZU-Vision*" or "*Recall*"), it is not because there isn't anything there, but it's been "blacked out." In more scientific language, this is called "*oc-*

clusion" —which is from the same root-word from which we get "occult," which simply means "hidden from sight."

We like to say the phrase "out of sight is out of mind" —but unfortunately that is not the case, that is just how an *Alpha-Spirit* got a better *game*. These old programmed tendencies continue to carry with us as we take on more, and hence the continuing *condensation* and *solidification* of *Existence*. This next exercise is meant to provide better *recognition* or *control* of this *"blackness"* phenomenon by handling it *knowingly*.

• Eyes open: *Look* around the *room*; *Spot* an object; alternately (using *intention*) *push* a *"black energy-wave"* over the object (concealing it from your view) and *pull* the *wave* off it. Do this repeatedly a few times; then repeat this on a different object.

—Eyes closed; using the previous instructions: this time *Imagine* a large object, building or even an entire cityscape (or landscape); then *push* and *pull* the *wave* to conceal it from your view. Do this repeatedly for a single object/scene, then repeat it on another (preferably larger-scale) scene.

The final exercise for this lesson is directed at improving the quality of *"mental imagery"* that a *Seeker* can *create/visualize*. This is another area we introduce *after Beta-Defragmentation*—and for good reason. While still in a *highly fragmented* state, an individual has very little control over their *"reactive pictures"* and *"mental reactivity,"* so any premature efforts to make the quality of that *imagery* stronger also risks essentially making an individual worse. Obviously we don't want that to happen.

As an individual gains greater control over their *compulsively created machinery*, they can obviously do more with it *knowingly*, and perhaps even reach a point to dispense with it altogether (knowing full well that they could create it again at will). What we want to *exercise* in this technique is a *Seeker's ability* to literally *Imagine* an *"Infinity"* —and even demonstrate that it's not difficult to *conceive* of.

• *Imagine* (*create/visualize*) a light-gray *road* just suspended in empty space. From a *viewpoint* above it: *Look* down its *"length"* in one direction and *consider* that it is of *"infinite length"* and goes on forever. Move your *viewpoint* to *Look* in the other direction and extend that *"length" infinitely* that way too.

–Now (looking toward one of the directions) add (*create*) a *mile-marker* or *sign-post* that says "0" on it, and place it beside the *road*. Move to a point you *consider* "one mile" down the *road* and add a *sign* that says "1" on it. Go another "mile" and do the same (with a "2"). *Notice* that you could essentially keep on doing this "forever" given the *consideration* that the *road* is supposed to be *infinite*.

–Now add (*create*) a single *mental image* that displays these *sign-posts* on the side of the *road* extending out *infinitely*, making a *"postulate"* that each one is numbered "one higher" than the former. [The *signs* can all be the same, stretching out in the distance; you don't actually have to be able to see all the "numbers" at the same time.]

–Now *move to mile-marker "100"* and see it there. Then: *move to mile-marker "12000"* and see it there. Then: *move to mile-marker "6350"* and see it there. Finally: *move to mile-marker "0"* and see it there. Practice alternating a *move* (in *perspective/viewpoint*) to each of these different locations on the *road* and *notice* the numbered *sign-posts* as you do.

NOTE: If a *Seeker* finds difficulty with this: refer to the previous step; repeat the *"postulate"* (*intention* that something will *Be*) that there are *"infinite sign-posts"* until it "sticks" (so to speak).

–Try to end the exercise at *"mile-marker 0."* Then throw the whole scenery away; and repeat the exercise.

We will leave any *end-realizations* regarding this practice for a *Seeker* to discover for themselves. If something does *"resurface"* in the process; simply refer to the appropriate *lesson* from the *Professional Course* and handle it accordingly. By this point of the *Pathway*, you're farther than you think.

LESSON FIFTEEN
"THE ARCS OF INFINITY"

SPHERES OF EXISTENCE

Here in the second-to-last lesson of the *Professional Course*, we move quickly through some more *advanced* concepts regarding the *Human Condition*, the *Game of Life*, and *Universes*. The lesson continues with areas treated previously. They are *"advanced"*; not because they are necessarily *"complex"* or difficult to understand, but because a *Seeker* is likely to have a greater appreciation for, or understanding of, them now—and how they might apply to one's own future progress on the *Pathway*.

Let's begin with the fact that we have no actual concern or question of the ongoing "existence" of an *Alpha-Spirit*. The *Spirit*, is as they say, *"Eternal"* or *"Immortal."* It is the *"conditions"* that one considers *as existence* that seem to decline and dwindle with each more *fragmented Universe* and each more *fragmented Lifetime*.

The *implanted-goal* of *this* present *Game-Universe* is *"To Survive"*— but that doesn't mean there is nothing else for an individual to pursue; it means that our *participation* in, and *experience* of, this level of *game* requires us to simultaneously apply *efforts* toward actively maintaining the *survival* of an *"organic game-piece."* It is from *this* perspective (in *Beta-Existence*) that we consider the *"Spheres of Existence"* (as first introduced in the *"Fundamentals of Systemology"* Basic Course).

From our perspective as a *Human Body* (or as a *lifeform* on *Earth*), the *Spheres of Existence* represent the various existing *fields* or *game-boards* that "spread out" in *Beta-Existence* from the *"Body."* We typically label the *First Sphere*, "SELF" (again, for perspective) —but what it really means for *this* game of *Beta-Existence* is: *surviv-*

al through a *Body* as an individual *"player."* It is only *as a "Body"* that we have any *"need to survive."*

The other *Spheres* are all *"aids"* or *"to one's own individual surviv-al"*; yet those that we consider as living their lives "seemingly oblivious" to others, would not have much *Awareness* extending beyond the *First Sphere.* It is easy to see, however, just how important these other *Spheres* are, as *"dynamic systems"* aiding *survival* of a *Body* (or individual).

1: PHYSICAL BODY; to meet the organic needs of a living organism.

2: HOME; to meet the domestic needs; housing, intimate relationships, sexuality, and procreation (physical survival continuing by future generations).

3: GROUPS; to meet social needs and participation with others (toward goals); includes jobs, clubs, and organizations.

4: SPECIES; to *exist/survive* as the *Human* species collectively.

5: LIFE ON EARTH; *survival* as a *lifeform* on *Earth*; *survival of* the *Earth*; includes *all living beings* on the *planet*, such as trees, animals and ecosystems.

6: PHYSICAL UNIVERSE; *survival* as a *lifeform* in *Beta-Existence*; *survival of* the *Physical Universe*; includes *all* the *space-time, energy-matter*, and *organic-life* of *this Physical Universe*.

7: METAPHYSICAL; the higher-level (*Alpha*) non-physical (often less obvious) influences of *Beta-Existence*; often limited to considerations of *"thought,"* but actually includes *all entities* and *spiritual fragments*.

8: INFINITY; is always *Infinity*, the ultimate *Alpha source*—but from the perspective of *Beta-Existence*, it tends to represent *"God"* or *"Divinity"* as the *source-point* of *Beta-Existence*; hence *Humans* tend to only approach this from the perspective of *"religion"* and *"worship."*

Before looking at this further, let's do some *light processing*. For this, you will simply *list* your *considerations* in your *notebook* (or

Flight-Log). It should be understood by this point of the whole course that we *process* our *considerations systematically* in order to *"free"* them from any particular *fixation* or *limitation*. We aren't *running* an idea or concept to the point of *apathy* (*"unfeelingness"*); we bring *Self-Determination* about those *considerations* back under greater control of *Self* by treating them *knowingly* (with *Actualized Awareness*).

The following *processing command-lines* (*"PCL"*) use the word *"could"* so that we are not restricting the *considerations* to what has/hasn't or is/isn't happening, *&tc*. We simply want a *Seeker* to explore these areas. [For present purposes: by *"survival,"* we mean the *continuance* of *game-experience* in *Beta-Existence*.]

A. *"How could a 'body' aid your survival?"*

B. *"How could you aid the survival of a 'body'?"*

A. *"How could 'family' or 'partners' aid your survival?"*

B. *"How could you aid the survival of your 'family' or 'partners'?"*

A. *"How could a 'group' you are a part of aid your survival?"*

B. *"How could you aid the survival of a 'group' you are a part of?"*

A. *"How could 'human society' aid your survival?"*

B. *"How could you aid the survival of 'human society'?"*

A. *"How could 'lifeforms on earth' aid your survival?"*

B. *"How could you aid the survival of 'lifeforms on earth'?"*

A. *"How could 'this physical universe' aid your survival?"*

B. *"How could you aid the survival of 'this physical universe'?"*

A. *"How could 'spirits' (or 'entities') aid your survival?"*

B. *"How could you aid the survival of 'spirits' (or 'entities')?"*

A. *"How could an 'infinity of creation' aid your survival?"*

B. *"How could you aid the survival of an 'infinity of creation'?"*

INFINITY & BEYOND

The *Spheres of Existence* are *systematically designed* and they relate to the other *seven-plus-one* models we use to gain a better understanding of our *Systemology*. In many ways, the *Spheres* are directly or observably demonstrable as progressively *larger* or *widening* "*spheres*" that encompass the *Body/individual* by necessity.

If you eliminate any *one* from the "*equation*" then the *lower* ones are also removed from *existence*. Therefore, having a *Body* (maintaining its optimum *survival* in *Beta-Existence*) is dependent on the *existence* of the *higher spheres* (hence "*Spheres of Existence*"). If you eliminate "*humans*" as a *species*, you eliminate "*human groups*" and obviously any individual "*humans*." Too greatly disrupting the ability to sustain an ecosystem for "*all lifeforms on earth*" will also dampen "*human*" existence; as is progressively being felt today on *Earth*.

Our purpose in emphasizing the "*Spheres*" is to enable a *Seeker* to *play* the "*Game*" here better. Of course, beyond this, as an *Actualized Alpha-Spirit*, our hope is that they would eventually "*ascend*" — rise above the entrapment within *this Existence* and be able to *create* a better *game* elsewhere — in an even higher *Universe*.

When we approach "*Infinity*" from the perspective of the *Human Condition* in *Beta-Existence*, we find the *Spheres of Existence*. Regardless of how someone might label or classify them, they are *systematically exactly* as described in our model. But as a *Seeker* reaches further on the *Pathway*, they find that the *Alpha* states and *Infinity* actually encompass more than what is immediately perceived at the start of the *journey*.

A *Seeker* will be the first to *realize* that "*To Survive*" is not the *only* possible *game* that an *Alpha-Spirit* is capable of. In fact, taking the whole *Backtrack* into *consideration*, it is probably the lowest-rung of potential *game-goals* we've experienced in our long existence.

When we discover the original native purpose or goal of an *Alpha-Spirit*, we realize that our preferred activity is *"To Create."*

Infinity is the "destination point" on the "horizon" that we use to align the direction of the *Pathway*. Of course, *Infinity* is always *Infinity*. But, from a *Human*, *"mortal"* or *"Beta"* perspective, *Alpha* states and *Infinity* are practically everything that is outside, or *exterior to, Beta-Existence*. The same model-style of *"spheres within spheres"* is often also used to represent *"Universes within Universes."*

When we break free of our *reality-agreements* for *this Physical Universe*, we do not suddenly arrive at *"Infinity"* in a literal or absolute sense; we arrive at a *higher-level Universe* with a *higher-level game-goal*, from which our *"Pathway-to-Infinity"* still continues. This reaffirms that *everything* that is beyond *Beta-Existence* is essentially *Alpha* by relative comparison.

When the *Alpha-Spirit* operates a *viewpoint* from within the next highest *"Universe,"* then *that Universe* is perceived as *"Beta"* relative to whatever *exceeds*, or *is exterior to*, it. A *Seeker* would then have to take what they achieved already from the *Pathway* in *the Physical Universe* to go further with it *there*—likely having to apply a similar regimen of *defragmentation* from the *viewpoint* of an *Alpha-Spirit* while *in* the *Magic Universe*.

THE ARCS OF INFINITY

Some *Alpha* qualities may be *systematically* worked out—using our models and lore—while still in *Beta-Existence*. This is because we know that the basic structure repeats itself at a higher-level—meaning, from our perspective, a higher *"harmonic"* or *"octave."* The *"dynamics"* represented by eight *Spheres of Existence* repeat from an *Alpha* perspective, with qualities that are not inherently within *the Physical Universe*. They cannot be appropriately called "*spheres.*" Since they extend beyond (or encompass) what we consider *Infinity* from a *Beta* perspective, we refer to them as the eight "*Arcs of Infinity*" for our *Systemology.*

9. (Arc 1): Ethics

10. (Arc 2): Aesthetics/Beauty

11. (Arc 3): Construction/Building

12. (Arc 4): Reason/Logic

13. (Arc 5): Variety/Randomness

14. (Arc 6): Games/Universes

15. (Arc 7): Understanding

16. (Arc 8): Creation

We *consider* the "*Arcs of Infinity*" as a *higher* "*harmonic*" or "*octave*" of the *Spheres of Existence* because, for example: "ETHICS" regards the "optimum actions" that enhance the likelihood of *survival as a* "PHYSICAL BODY." The idea and reasoning behind "*ethical action*" is not inherently a part of *this Physical Universe.* Those particular "thoughts"—or *intentions for action*—are not found within the motor-functions or "brain" of a *Body*, which when left to itself operates solely on *stimulus-response.* Those *higher considerations* are "*Alpha*" qualities because they *originate* from outside of, *or exterior to*, this *Beta-Existence.*

When we think of the *Second Sphere* of "HOME" as including loving and close intimate relationships (which are more "emotional" in nature), it is not difficult to see its connection to *game-goals* of the former *Magic Universe* ("*To Enjoy*"), which emphasized sensual pleasure. The *Second Arc* is "AESTHETICS" (which is also to say "beauty")—and here again we see a domain of "*thought*" or *consideration* that originates beyond what is contained exclusively within *the Physical Universe.*

By themselves, the *Spheres of Existence* and *Arcs of Infinity* are simply the basic structure of a *game* that we may have once even participated in *creating.* This alone is not the source of our *frag-*

mentation. But these *games* did, and do, provide the *foundations* or *platforms* by which additional *implanting* and *programming* is laid in—and upon which even more *imprinting* and *fragmented considerations* compound on top of (or attach to). This means that many "positive" *goals* and *expressions* have only become *fragmented* by being used negatively.

Use the labels given for each of the *Arcs of Infinity* and insert them (one at a time) into the PCL-*formula* that follows:

A. *"How could your ---- enhance the survival of others?"*

B. *"How could someone else's ---- enhance your survival?"*

By *defragmenting* our *considerations* and regaining *spiritual abilities* of these *"higher"* aspects (or *Arcs*) in *this lifetime*, we not only stabilize and increase our own progress on the *Pathway*, but also *enhance* the *conditions* of *survival* while still in *this Game*. By enhancing them and ceasing to use them against one another *first*, we "unlock" or "unfold" more of our potential *thereafter*. But the *"livingness"* that we experience becomes more optimal all along. We do not have to *"suffer"* in *this Life* in order to have some better tomorrow "somewhere else." [That particular idea is just one example of *"implanting"* on a *"religious"* level.]

Use the labels again for this PCL-*formula*:

A. *"How might making a choice based on increasing ---- lead to an optimum action?"*

B. *"How might making a choice based on increasing ---- lead to an undesirable (or harmful) action?"*

By *"optimum action"* we are indirectly referencing *Ethics*, which is the *First Arc*. The subjects of *Ethics* and *Aesthetics* are probably easiest to handle in a single lesson for *Systemology Level-6*. Many of the other areas are introduced throughout the *Professional Course*. *Ethics* and *Aesthetics* also most closely relate to *this Physical Universe* and the *Magic Universe*, which are the two main *"Universes"* we've covered on this course.

"COURAGE" & "BEAUTY"

Let's begin by *inspecting* some of the qualities inherent in the *lower* *"Arcs."* For example: our conception of *"courage"* became *fragmented* early on the *Backtrack* by glorifying how *beautiful* it was to be *courageous*, which—for the *Games-Universes*—meant to *"fight against overwhelming odds and lose."* [Because as we've seen in former lessons, these *Universal-Games* are designed so that everyone loses.]

At these *advanced processing-levels*, we take some of the "charge" off of the *effects* from these early obscure *implants* by *"Imagining"* a *facsimile-copy* (or an *approximation*) of whatever we are treating—or by *"Getting a Sense"* of some concept *real* enough to actually *process* it. Some of this is a far-reach in ability for *Seekers* early on, which is why we prefer to handle *"Conceptual Processing"* after effectively completing lower *processing-levels*.

In this case, let us introduce one example for *processing-out* the *"postulate"* (or *"Alpha-Thought"*) that: *"one's opponents are stronger when one is being courageous."* *Run* the following PCL in series repeatedly until you feel no *inclination* or *tendency* to want to *"make an opponent stronger to show off how courageous you are."*

A. *"Get a sense of the beauty of being courageous and losing."*

B. *"Get a sense of the beauty of someone else being courageous and losing."*

C. *"Get a sense of the beauty of being courageous and winning."*

D. *"Get a sense of the beauty of someone else being courageous and winning."*

Now: *Imagine* the *"feeling"* of *being strong* and *courageous* without the need to face any opponents in order to prove anything. [If this seems difficult, there may be more *"charge"* on the area, which means *running* the previous PCL-series longer.] The purpose is to regain and retain the actual characteristic of *"courage"* —which is technically, *"the ability to confront or willingly face anything"* — without the *fragmentation* attached.

Once this is accomplished: using your experience with techniques from earlier lessons; *Imagine clouds* of this *clear/pure/true courage* all around you, and absorb them into the *Body.* Additionally, practice doing this around others, or groups of people, as a *cloud* above and/or surrounding them. In *Zu-Vision: Imagine* these *clouds* above large cities, countrysides, nations and even the entire planet.

Although these exercises employ *creations* that only *you* are *agreeing to* as a *reality*: if the true and pure *intention* (of *"courage"*) has been *realized* (or *"contacted"*) and is being used here, then it *is* an effective *process* for, at the very least, your own benefit (or progress). This is one way of handling the *Arcs* for this course.

Often the idea (or *concept*) of *"beauty"* is also attached to "pristine states" or "newness." One would note this to be in effect, at a physical level, regarding our *perception* of the *"young."* One way this is handled in *processing* is to *run* the idea (or *concept*) of *"decay"* and gradually increase a *Seeker's* "acceptance levels" of it.

For example: *Look* around the *room*; *Spot* an *object*; and *Imagine* a *facsimile-copy* of it. Throw a few away and keep at it until you are satisfied with one (rather than trying to make an existing one more vivid). Then: *Imagine* (*visualize*) your *copy* as decaying (rusting, collapsing, deteriorating, *&tc.*) until it breaks apart completely into a bunch of shattered *fragments.* Then: *run* this in reverse; having the *fragments* come back together (and "heal") until it matches the original state again. Alternate this several times on one *object*; then select a different *object.* This is *run* until a *Seeker* can *confront "decay."*

To advance this further: a *Seeker* applies the above steps conceptually to *living beings* (trees, animals, other people); starting with those that can be *viewed* physically, and then using *imagination* (or *ZU-Vision*) to conceive of *living beings decayed* that are not physically present. This can be applied to *larger systems*, too—*confronting* the *"decay"* of cities, civilizations, planets, galaxies and even *Universes*.

An *end-goal* of *Systemology Level-6* is for a *Seeker* to be able to *perceive, think,* and *evaluate "freely"* on a moment-to-moment basis, in their *present-time* experience of *Existence*. *Beta-Defragmentation* does not automatically *"release"* one from *this Game-Universe*; but a *Seeker* experiences a sense of *"release"* while *in this Game*, when they can fully recognize that they do not actually need to *fight for survival*—at which point much of the other *turbulence*, and struggles with remaining *fragmented* areas, just starts to *"fall away."*

THE ARC OF "AESTHETICS"

Aesthetics is an *Alpha*-quality *consideration* about *how* something *is* or *appears to be*. It is not enough to simply call it *"beauty,"* although that is one of its areas or aspects. Our conception of what is *"beautiful"* compared to what is *"repulsive"* is a *reactive quality* contained in most *fragmentation* in *this Universe* (and also, at the very least, the *Magic Universe*).

We tend to consider *"aesthetics"* and *"art"* to be *"positive"* expressions of *creative ability*; and they *are*—when they are not *fragmented*, or blatantly used *against others*. *Beauty/Ugliness* can quite easily enhance the *fragmentation* connected to *reactive "reach"* and *"withdrawal."* We can be easily *compelled* to interact, *communicate*, *agree, &tc.* with those that we find *"attractive"* (for whatever underlying reasons we might make that determination), even if it is not in our *"best"* interests. *Politics* (*marketing* and *social opinion*) can be won this way; fidelity of strong relationships can disintegrate under its effects; the list goes on.

As cliché as it has become among the speech of independent artists: the first taste of this *Arc* is often glimpsed when one is truly "taken aback" or "fascinated by" the sudden impact of true *"beauty in the world."* This is something that is not innate to primitive survival and is not subject to reasoning or critical thought. This is one reason *aesthetics* remains an *"Alpha"* quality; because there is no equivalent "program" for it built into the systems of *the Physical Universe. Aesthetics* is the very thing that differentiates what is *"beautiful"* from an *"ordinary mundane creation."* And *that determination* can only come from an *Alpha-Spirit.*

For a *Seeker* to *analyze* any such *considerations,* use the following PCL; list readily available responses ("B") and then repeat "A" with something different. All we are doing here is bringing some things *up* to examine.

A. *"Recall something aesthetic (or beautiful)."*

B. *"Spot something about it that contributes to the aesthetic/beauty."*

While elements specific to *this Physical Universe* might seem inconsequential at *higher-levels*—such as to a *spiritual entity outside of it* at the *Seventh Sphere of Existence*—the *Arcs of Infinity* are introduced at the end of the *Professional Course* as suggestions for further *Alpha-Defragmentation* (which is the next main focus on the *Pathway* once a *Seeker* has successfully completed the work described in the *Professional Course* lessons).

Aesthetics have even more impact (or "power") at a *Spiritual/Alpha* level. In an *existence* where the higher *goal* is *"To Create,"* aesthetics are a major component of getting others to accept one's *creations,* or when trying to *create effects* on others. So, there is a lot of jealousy, manipulation, and critical evaluation attached to our experience of *aesthetics.* Here, a *Seeker* can *run* the following PCL to *confront* some of the *"charge"* in this area.

AESTHETICS (ENFORCED)

1. *"Recall (en)forcing an aesthetic on another."*

2. *"Recall someone (en)forcing an aesthetic on you."*

3. *"Recall someone (en)forcing an aesthetic on another."*

0. *"Recall (en)forcing an aesthetic on yourself."*

AESTHETICS (INHIBITED)

1. *"Recall inhibiting an aesthetic for another."*

2. *"Recall someone inhibiting an aesthetic for you."*

3. *"Recall someone inhibiting an aesthetic for another."*

0. *"Recall inhibiting an aesthetic for yourself."*

And then balance the above by *running* "positive action" (below) to end this cycle of *processing.*

1. *"Recall someone enjoying an aesthetic that you shared with them."*

2. *"Recall enjoying an aesthetic that someone shared with you."*

3. *"Recall someone enjoying an aesthetic that another shared with them."*

0. *"Recall enjoying an aesthetic you discovered for yourself."*

As before with *"enforcement/inhibition,"* we will *run "evaluation/invalidation"* on this area; but instead of treating an *aesthetic-item* itself, this *process* pertains to *considerations* (*opinions, &tc.*) of what *is aesthetic* (or *beautiful*). [*"Evaluating"* is giving the *"opinion"*; and *"Invalidating"* is negating an *"opinion."*]

AESTHETICS (EVALUATED)

1. *"Recall (en)forcing an evaluation of an aesthetic on another."*

2. *"Recall someone (en)forcing an evaluation of an aesthetic on you."*

3. *"Recall someone (en)forcing an evaluation of an aesthetic on another."*

0. *"Recall (en)forcing an evaluation of an aesthetic on yourself."*

AESTHETICS (INVALIDATED)

1. *"Recall invalidating an aesthetic for another."*

2. *"Recall someone invalidating an aesthetic for you."*

3. *"Recall someone invalidating an aesthetic for another."*

0. *"Recall invalidating an aesthetic for yourself."*

Then, as before, we balance this type of emphasis by *running* some positive action, such as *encouragement.*

1. *"Recall encouraging someone's creation of aesthetics."*

2. *"Recall someone encouraging your creation of aesthetics."*

3. *"Recall someone encouraging another's creation of aesthetics."*

0. *"Recall encouraging yourself to create aesthetics."*

In practical application: the first step to any area of *creation* is simply *"To Create."* This means *being creative* without the concern of *aesthetics*, *quality* or *content*. Just as one's ability to *knowingly create "mental images"* improves with practice, so do other areas. For example: if you want to "write," then just start writing a lot, for yourself, until you find it easy to sit and write. Afterward, you can begin to work on *creating writing* that *others* would want to read. In other words: start with the *creative ability*, then add the *aesthetics*. This can really be applied to any "artistic" or "creative" endeavor.

BASIC ALPHA-DEFRAGMENTATION

A *Seeker's* starting point for *"Alpha-Defragmentation"* is to be found with the *Arcs of Infinity*. Obviously, this lesson is only the beginning to a full handling of these areas. These areas extend much further on the *Backtrack* than what we even have to deal with concerning the *"Physical Body"* and this *"Beta-Existence."* This also means that as a *Seeker* advances solidly on the *Pathway*, more will become *accessible* for *inspection* because an individual is *willing* to *confront* more and more of what has been overlooked or forgotten.

Since we have already examined *Ethics* and *Aesthetics* to some extent, let's take a look at the other *Arcs*. In traditional *defragmentation processing*, we avoid the word *"think"* in our PCL to avoid triggering or activating unnecessary *mental machinery* and *circuitry*. *Systematic Processing* is focused and targeted; it doesn't involve a lot of "freewheeling thought" if ting unnecessary *mental machinery* and *circuitry*. *Systematic Processing* is focused and targeted; it doesn't involve a lot of "freewheeling thought" if the *processes* are to be effective. However, the *processes* below are basic explorations into these *Arcs*—simply permitting a *Seeker* to get a view of each area.

A. *"Think of something you would have fun building."*

B. *"Think of something someone else would have fun building."*

A. *"Think of an interesting logic-puzzle you would have fun solving."*

B. *"Think of an interesting logic-puzzle someone else would have fun solving."*

A. *"Think of some interesting changes that would be fun to have."*

B. *"Think of some interesting changes that would be fun for another."*

A. *"Think of some games that you would have fun playing."*

B. *"Think of some games that someone else would have fun playing."*

A. *"Think of something that you would find interesting to understand."*

B. *"Think of something that someone else would find interesting to understand.*

A. *"Think of something that would be interesting to create."*

B. *"Think of something that someone else would find interesting to create."*

While it is easy to see the "positive characteristics" inherent within these *Arcs*, there is unfortunately the *other side* of them for us also to *confront*. We have *all* used these *Arcs* against each other, at one time or another, on the *Backtrack*. The *lower-level control* mechanisms—like *"pain"*—really have no effect on a *godlike being*. But the *Arcs*, on the other hand—*ethics, aesthetics, building, games, &tc.* —are of extreme interest to *godlike beings*.

All of the *Arcs*—not just *"aesthetics/beauty"*—can be, and have been, used as *"lures"* for entrapment. And early on the *Backtrack*, the *Alpha-Spirit* is quite *"innocent"* and can be easily *"tricked."* These *Arcs* serve as the desirable *"bait"* used for laying in *higher-level fragmentation*—*"Spirit-Traps"* and *"Implants."* Therefore, by properly understanding them, additional data for many of the *"Implant-Platforms"* (used in *advanced processing*) are derived.

It is important to keep in perspective that we have all been participants in these areas; we have experienced both sides—and so these areas are not to be treated as "one-sided" ordeals. A *Seeker* should be prepared for *confronting* anything that "comes up" in *processing;* and even at *Systemology Level-6,* only some of the *most accessible* content (or *"almost knowns"*) will be available for *processing* on the first pass through the course.

PCL for these *advanced processes* (below) employ the phrase *"Spot the impulse."* This is meant to result in the *recall* of a particular *"consideration"* or *"intention,"* but can also regard a specific *"incident."* In the absence of anything accessible: a *Seeker* may choose to "soften up" an area first by using *"Get a sense of"* as a *resurfacing* technique. The basic *"Hot Button"* (or *"keyword"*) we will use for these *processes* is *"control (or manipulate)."* These *processes* use high-level *Awareness* to *defragment* by *inspecting* some tendency *"As-It-Is."*

A. *"Spot the impulse to control (or manipulate) others through their sense of ethics."*

B. *"Spot the impulse of others to control (or manipulate) you through your sense of ethics."*

A. *"Spot the impulse to control (or manipulate) others through their desire for beauty (or aesthetics)."*

B. *"Spot the impulse of others to control (or manipulate) you through your desire for beauty (or aesthetics)."*

In this next *Arc,* "construction" could be building material structures and things, or it could be building an organization and group. A *"Free Being"* will often *join a group* to accomplish a large constructive activity.

A. *"Spot the impulse to control (or manipulate) others through their joy of constructive efforts."*

B. *"Spot the impulse of others to control (or manipulate) you through your joy of constructive efforts."*

"Logic" is an entire subject unto itself. It is also used to entrap by manipulating an *agreement* (to which a *Seeker* should also study *"logical fallacies"*). A *"logical argument"* is meant to persuade an individual that such-and-such (a *conclusion*) is the case only because certain other things (*premises*) are agreed with as true.

A. *"Spot the impulse to control (or manipulate) others through their agreements with logical arguments."*

B. *"Spot the impulse of others to control (or manipulate) you through your agreements with logical arguments."*

An *Alpha-Spirit* exists for a very long time and is fond of *variety* and *change*, which is to sometimes say *novelty.* If an individual can be convinced that "something else" is just as good (or better) than what they "have now," they will usually *change it* or *trade it* just for the "fun" or "novelty" of it. But sometimes an individual becomes highly resistant to all *change* on a *reactive-automated* basis simply because this "trick" has been played on them too many times.

A. *"Spot the impulse to control (or manipulate) others through their desire for change (and variety)."*

B. *"Spot the impulse of others to control (or manipulate) you through your desire for change (and variety)."*

A. *"Spot the impulse to control (or manipulate) others through their desire to have an interesting game."*

B. *"Spot the impulse of others to control (or manipulate) you through your desire to have an interesting game."*

An *Alpha-Spirit's* desire to *"understand"* can even be turned against them. Just one of the more recognizable manifestations of this is *"playing victim."* In this case, someone is trying to *convince* you that you can't really *understand* how they *feel* unless you *feel* just as *"bad"* as they are pretending to *feel.*

A. *"Spot the impulse to control (or manipulate) others through their desire to understand."*

B. *"Spot the impulse of others to control (or manipulate) you through your desire to understand."*

At the peak of all this is the *Arc of Creation* — which essentially also encompasses *all* of the others; just as *Infinity* encompasses *all* of the *Spheres of Existence*. The *Alpha-Spirit* loves *creations* and *creating* above all else. We *imagine* "things" and make *facsimile-copies* of "things" just to *have things*.

To some extent, the *Pathway* is graded by an increase of *creative ability* — and by this, we mean the *certainty* a *Seeker* has on *having things*. Otherwise, they will not "let go" of some *fragmentation*. At the root of it all, an *Alpha-Spirit* is afraid of not *having enough variety* and *richness* to their *creations*. As a result, we can be tricked into *copying* and *creating* "undesirable" *reality-agreements* just to keep a "library" of "interesting details."

A. *"Spot the impulse to control (or manipulate) others through their desire to create (or experience an interesting reality)."*

B. *"Spot the impulse of others to control (or manipulate) you through your desire to create (or experience an interesting reality)."*

Proper *defragmentation* of any area treated on the *Pathway* does not result in an *abandonment* or *withdrawal*. It results in rehabilitating the *true abilities* or *clear characteristics* by shedding the debris that enshrouds them. This is why we balance handling *deeper* or more *turbulent* areas with "*Spotting*" positive action. Rather than apply "*objective processing*" for these areas, a *Seeker* should look over the "*Arcs*" (list) and *knowingly* practice ways to *actually* develop in these areas during this lifetime, however trivial a particular activity may seem.

CREATING "INFINITIES"

At the end of the previous lesson, we presented an *advanced visualization* exercise for conceiving of "an infinite roadway." Here, we pick up from where we left off (and a *Seeker* may wish to review that material before continuing).

• *Imagine* (*create/visualize*) a "*grain of sand.*" *Look* at it closely; *Notice* some of its details (exact shape, texture, color).

–Now: *Imagine* a *dozen* different individual "*grains of sand*"; making each a slightly different shape, texture, or color.

–Now: *Imagine thousands* of *copies* of these different individual "*grains of sand,*" all mixed together forming a "*small area.*" You should be able to perceive a "*speckled-like variety*" in the scenery rather than simply an entire area of one fixed color, texture, or pattern.

• Building on what you practiced in the previous lesson: "*copy*" the "*small area*" (from the previous step) as many times as is necessary to get a sense of the "*sand*" stretching out like a large "*beach.*"

–Now (similar to how you had handled the "*road*"): *Imagine* (*create/intend*) that the "*beach*" extends out to "*infinity*" in both directions.

–Now (similar to how you had handled the "*road*"): *Move* to various "*locations*" on the "*beach*" to *get a sense* (or *certainty*) that it does continue as far as you are interested in checking it.

• *Look* around the *room*; *Spot* an *object*. *Imagine* (*create*) a *copy* of it, overlapping, but shifted just slightly to the side.

–*Imagine* (*create*) another *copy*, overlapping, but shifted slightly to the side of the last *copy*. Then: *make* many more *copies*, each one a little bit over, until they appear fanned out in a line, like "*cards.*"

–Now: *Make* as many *copies* of the "*object*" (or even the "*fanned out image*") as is necessary to *Imagine* a "*line*" of this *object* going on forever. When you've accomplished this to your satisfaction, "throw away" that *mental imagery*.

–Now, similar to the beginning step of this exercise: *Imagine* (*create*) a single *copy*; but this time, *Make* the *"infinite line"* of the *object* occur with one single *"intention"* (or *"postulate"*). If you have difficulties with this: *Imagine* the "coloration" gradually varying as the *copies* extend into the distance.

–Then *choose* a different *object* and repeat the *process*. Practice until you are comfortable *"creating infinities."*

Finally, the following *process* assists increasing *spiritual perception*. It is appropriate for *ending-sessions* with, and also as a stand-alone exercise.

• *Look* around the *room*; *Spot* precise *"points"* (or *"spots"*) in the *"space"* of the *room*. [By this we mean in *"mid-air"* and not connected to, or touching, an *"object."*] Really *get a sense* that they are there, even though they have no *"mass"* to them.

–When you are comfortable with the previous step: *Spot* specific *"points"* (or *"spots"*) that are *outside of* the *room*. [This can be done with eyes open or closed, but it is done mentally (or with *ZU-Vision*) rather than with the *Body's Eyes*.]

–Alternately: *"Spot three points in the room"* and *"Spot three points outside the room."*

–When you "feel good" about the exercise: *"Get a sense of the floor/ground beneath you"* and *end-session*.

LESSON SIXTEEN
"ALPHA THOUGHT"

QUANTUM SYSTEMOLOGY

Many years ago, *archaic* or *early* presentations of our *Systemology* borrowed more *terminology* from the preexisting material sciences. The *systematic vocabulary* we use today, and for this course, continued to evolve for over a decade among those *Seekers* participating in our underground development (since *2008*).

In its early development, we noticed that the *Seekers* catching on the quickest all had something in common: a previous conceptual understanding of *"metaphysics," "physics,"* and/or *"quantum physics."* There are actually many concepts relayed in *"quantum physics"* that demonstrate an evolution in understanding by *some* scientists regarding the *"physics"* of *this Physical Universe.*

Understanding the study and unique terminology of *quantum physics* is hardly necessary for the *Pathway.* However, an *advanced Seeker* may find a "conceptual" overview quite interesting. And it is always helpful to understand the actual *"physics"* of whatever *Universe* one is "operating" from. [Such an endeavor exceeds the scope of this lesson. We direct *students* to other material for this, such as Fred Alan Wolf's *"Taking the Quantum Leap: The New Physics for Non-Scientists."*]

One of the basic things a *Seeker* can take away from the idea of *quantum physics* is that the actual *reality* or *solidity* underlying what we observe as the *Physical Universe* is not quite as solidly fixed as conventional physics would have had us believe; that things are not all "clockwork gears" and "solid billiard balls" knocking into one another; and that an individual, themselves —*"The Observer,"* as they say—actually has some participation in the *reality* of things being how they are, seem, or appear. Of course, an *advanced "Systemologist" knows* all this from their own experience, without being concerned with formulas or math.

Our *Systemology* proposes an *Infinite "background"* to *ALL Existence* that is essentially a *Nothingness*—an *Infinity-of-Nothingness*. This is, however, balanced by an *"infinite potential"* of *Somethingness*. It doesn't mean that this *infinite-everything* is experienced or manifest at once; it means that the *"potential"* for something *TO BE* is actually *limitless*. Of course, this does not rule out bringing something into *being* that will, at a certain level beneath that, *limit* the *potential*.

For example: If you hold the *decision* that you "can go anywhere" at the same time as "cannot go on a plane," you've actually *limited* your *potential* of action (or freedom); and probably are continuously creating *fragmentation* about *"planes,"* which eventually can become as *solid* as a *barrier*. Right now, we aren't concerned with where the individual goes or how—all that is relevant here is the *layers of decision* that exist.

In *quantum physics*, there is a concept called a *"probability wave"* or *"potentiality wave function."* All this means is that *all* states of *"possible existence"* are suspended in a condition of *"potential"* until they are *realized*. The analogy we might cite of *"Schrodinger's Cat"* ranks right up there with *"Pavlov's Dog"* in demonstrating just how cruel and out of touch these "scientific minds" are with the natural world they claim to understand.

As a more civil example: if you flip a "coin" so that it lands *behind* you (out of view) and there is no one yet observing the *"heads/tails"*result, then the *potential observable reality* is *simultaneously both*. The *"wave"* of *potentiality* has not yet *"collapsed"* into a *"solid reality"* of one position or another until it is actually *observed*. At the point it is *observed*, only *one* of the possible *"positions"* or *"functions"* of the *"potentiality-wave"* becomes a *Something*; and the others fade back into *Nothingness*. This actually says a lot about our *Universe*.

While this level of participation does not *change*, for example, the molecular structure of the "coin" into something else, it demonstrates an underlying principle behind *reality*, which *"magicians"* and *"mystics"* have been attempting to *control* for a very long time —long before any *"quantum sciences"* were recognized. For the final lesson of this *Professional Course* we will approach these same areas, but *systemologically*.

THE "COIN" EXPERIMENT

Our "coin" example from the introduction was not chosen at random. It reflects a very real experiment that a *Seeker* can play with. The basic instruction is given in the example; but note that the act of *observing* the result *"collapses the potentiality wave"* and thereafter cannot be changed. So, anything a *Seeker* might attempt to *do* or *create* (*visualize*) must take place before the "coin" is examined to see if it shows *"heads"* or *"tails."*

"Quantum Effects" suggest that we can influence the *"probability wave"* before it *collapses* into one position as *reality*. You *"visualize"* or *"intend"* prior to *flip-tossing* the "coin" or afterward; but any influence that might be had will end as soon as it is observed by someone. You might try *visualizing* "coin come up heads" or perhaps *create* a large *mental image* of a "coin showing heads" in front of you. If you are practiced in *ZU-Vision*: you might *look* at the "coin" while it's still behind the *Body* (with eyes closed) and *create* a *copy* of it (showing heads) hovering over the physical coin. Any specific method requires personal experimentation.

As a systematic experiment, an individual will attempt to *run* this *10 times*: performing the *visualization* and then *observing* the *reality*. Record the results in your *notebook/log* of how many times it comes up *heads* versus *tails*. You might make additional notes about the technique you used (and any other relevant data). [As an experimental procedure: conduct a *"run of 10 times"* once daily for a week and examine the results.]

Of course, mathematically, the *coin* should come up *heads* about *5 times (of 10)*. Our data from *uninfluenced "runs of 10"* demonstrated that it is not uncommon for *heads* to "naturally" come up *4 to 6 times (of 10)*. We use *"runs of 10"* because it's easy to treat as a *"percentage"* and of course we do this experiment for several days to compare results. This helps you decide if you need to alter the technique(s) being applied.

A consistent *7 times* (70%) or *greater* (of *"visualizing" heads* and having *heads* coming up) is considered a possible *probability bias* in your favor. There are other *spiritual mechanisms* that can be at play here—including a tendency to "attract negatives" (where you get *tails because* you're trying to get *heads*). For example: if consistently coming up with *3 times* or less, you may still be influencing the *probability*, but in a way where you are *unknowingly* working against yourself. [If you are applying *intention* and coming up *3 times* or less, it is likely you still have some *Beta-Defragmentation* to do.]

This experiment is simply that; an experiment. It is not, by itself, any indicator of one's own *spiritual ability*, or even a checkpoint for the *Pathway*. It's simply something fun we added that is relevant for our lesson.

WHAT IS A "POSTULATE"?

Throughout this course (and other *Systemology* material), perhaps one of the most *esoteric* words used is: *"POSTULATE."* We tend to prefer our own term *"Alpha-Thought,"* though they really mean the same thing when treated at the highest level of *Beingness* and *Existence*. But, of course, we have merely exchanged one word for another. So, what does it actually mean? What *is* a *"postulate"*?

A specific *systematic* and *scientific* use of the word *"postulate"* began over *2000* years ago in ancient *Greece*, when the mathematician *Euclid* started systematizing an understanding of *"geometry"* for the *Greeks*. In order to do this, *Euclid's* demonstrations began with *"first principles"* that must be taken as absolute without proof. While the *Greeks* already had the word *"axiom,"* it wasn't quite right for this. His *"axioms"* were *true* only when based on other even more "basic" statements of *truth*, which he called *"postulates."*

The word *"postulate"* comes from the *Latin "postulare,"* meaning *"to demand."* A *postulate* stands on its own; it does not require *logic* or *reason* to *Be*; and in *Euclid's* case, truths that required no proof. From these basic truths, other *axioms* can then be established—using the *postulates* as the basis for *their logic* and *reason*. This is where we get ideas such as "a line segment connects any two points" and "parallel lines never intersect."

Where it concerns our *Systemology* of the *Alpha-Spirit*, we treat *postulates* as *"Alpha-Thought,"* which is to say *"causative thought."* This is treated at "7" on our *Standard Model* and *ZU-Line*, the same position that we place the *Alpha-Spirit* on our charts. This is because at the level of *postulates* or *Alpha-Thought*, we are directly treating the *native state* of the *Alpha-Spirit* itself. When we *Imagine* or *Create* a *mental image*, we are directly *"postulating"* it *to exist* within our *"Personal Universe."*

Alpha-Thought is the *origin-point* of everything that is directed from the *Alpha-Spirit* all "down" the *ZU-Line*. But our *Awareness* and *intentions* must also pass through various levels of *"implants"* and *"mental machinery"* before resulting in what we consider our *experience* of *Beta-Existence*. This is why *systematic processing* is so critical for rehabilitating *Alpha-Thought*; because our full *Self-Determinism* is being *limited* by *fragmentation*.

We have indirectly treated personal *postulates* for the entirety of the *Pathway*. It is the level of *causation* that permits an individual the *freedom* to *actually change their mind* or even *create*. The various exercises and *processes* strip away the layers of *restriction* that remain from former *postulates* and *creations* we've forgotten.

Postulates are a very specific kind of "thought" that we carry around; but it is not a "thinkingness." It is a direct statement or *intention* to *Be* or *do*, or even *Not-Be* or *avoid-doing*. It may be an intention to *have* or *not-have*. The *justification-considerations* we use to make others wrong, or as reasons for our failures, are usually *postulates*. And *postulates* sometimes include absolutes like *"always"* or *"never"* —such as "I never do things right" or "I always do things wrong." The statements usually will have some *turbulent charge* on them.

Many *postulates* will include *"I," "I am,"* or *"me,"* in its wording. They are typically stated in present-time (not the future). "This is killing me" or "This kills me" is a *postulate*. "This might kill me one day" is *not*. A child trying to get out of attending school by saying "I am sick" is making a *postulate*. Now, that *postulate* may not actually *"stick"* and be *reality* at that time—but, enough years of repetitively putting *"charge"* on that statement *could* result in a manifestation at any point; because *Self* does not like to be *wrong*.

In *Traditional Piloting*, a *Professional Pilot* will record the *postulates* that a *Seeker* says while *incident running* or other *systematic processes*. These are *run* after *confronting* an *incident*, *&tc*. This is why we instruct a *Solo-Pilot* to notice any *decisions* they made during, or as a result of, an *incident*. These *postulates* are what hold the *fragmentation* in place, and ultimately is the area we want to treat for effective and stable *defragmentation*. Most of the time, we are simply reducing the *charge* or *turbulence* in an area in order to make *confronting* things more acceptable; but *total defragmentation* also requires handling an underlying *postulate "As-It-Is."*

CAUSATIVE THOUGHT

The original *native state* of an *Alpha-Spirit* includes near-infinite *creative ability*. This is what the *Alpha-Spirit* enjoyed in its highest purest form in its own *"Home Universe"* —and is mirrored in our

"Personal Universe." At this ultimate or highest level of *spiritual ability*, the *Alpha-Spirit* is able to have "things" *exist* by simply *postulating* that they *exist*—or having things "happen" simply by *postulating* that they *are happening.*

This *"Alpha"* level of *causative thought* is what *magicians* and *New Thought practitioners* are chasing after, each in their own way. *Alpha-Thought* is not an internal activity, such as "wishing," "hoping," "praying," or "desiring" strongly for something to *Be* or *happen. Alpha-Thought* is an "external projection" of *thought* that "demands" manifestation in *reality*; even if that *reality* is only one's own *Universe.* Of course, there are many things we *intend* or *create* that *do* actually manifest in *Beta-Existence.* But, it is easiest to make your *postulates* work (or *"stick"*) when treating *creations knowingly* in your *Personal Universe.*

For example: *"Visualize a door."* That "door" is there; it *exists* in your *"imagination"* or *Personal Universe.* But, this is an example of a *postulate* or *Alpha-Thought.* It is not a *"request"* for the "door" to *appear*—or any application of *"effort"* in order to *create* it. *"I have a door"* or *"having a door"* is the *postulate.*

Now: *"Visualize the door opening."* Again, a *postulate*; it is *your* "door" in *your created space*; so it just *does.* Of course, even with one's own *creations*, if you were to just sit there and "look" at the "door," or even *wish* about what it should do, it wouldn't do anything. If a *Seeker* does have to start off thinking of *postulates* as similar to *wishing* (for internal dialogue or something): they can "wish" to *have* the "door" open; but then *have* the "door" open. The actual *"having it open"* is the *postulate.*

Making your *postulates* work (or *"stick"*) in *the Physical Universe* is a bit more difficult. For one: it is a *Shared-Universe* that is held together by *reality-agreements* of many individuals, with each having their own *Personal Universe.* Each interacting individual is maintaining some *agreement* with *this Universe*; but they are also *postulating* things, maintaining *compulsive-continuous creations,* holding onto *counter-intentions*—albeit *unknowingly*. Most individuals on this planet are in some way affecting other individuals on this planet. And we might even say *"fragmentation is contagious."* But, highly *Actualized* individuals—able to *confront* the *games* of this existence—also have a uniquely strong ability to positively affect the world around them too.

At this stage of the *game*, an *Alpha-Spirit* is so deeply enmeshed in *reality-agreements* with *the Physical Universe* that it is rather difficult to make *Alpha-Thought* manifest directly in *Beta-Existence*. By *Alpha-Thought* we mean *intentions* that *materialize* while bypassing Physical Laws; or in other words *"magic."*

One's own *considerations* and *counter-intentions*, not to mention out-of-sight *spiritual machinery*, all contribute to *"dampening"* or *"limiting"* direct material manifestations of our *spiritual ability*. And by this, we do *not* necessarily mean literal demonstrations of "pulling rabbits from thin air" or "making solid forms disappear." But we *do* mean the kinds of things that individuals are likely to *"wish"* about, or even *"cast spells"* for, *&tc.*

Unlike other approaches to "personal development," our *Systemology* emphasizes *defragmentation* rather than starting off with attempts to manifest *causative thought*—such as you find with typical *New Age* and *New Thought* traditions. Much like the handling of *"energy"* and *"energy beams"* (in former lessons), there is a lot of potential *invalidation* to overcome when applying techniques directly to *the Physical Universe*. When this happens too much, it tends to reinforce a conviction that *"Beta-Existence* is *right;* and we are *wrong."*

The easiest *postulates* to make "stick" are those you make for the benefit of others rather than yourself. You could still directly benefit from this, if you are able to *postulate* things for others in a way that also brings you what you need as a byproduct. But really, don't get too frustrated if you don't get immediate results.

Our entire *Systemology* is in many ways a study (and handling) of the *considerations* and *counter-intentions* that block our *Alpha-Thought* (or *spiritual ability*) from being as high-powered as we can be in our *native state*. Your *native state* is not "lost" to you—because it *is you*; simply "hidden" by *fragmentation*. And in *our tradition*, this progressive uncovering and clearing of *Self* is what we treat as the *Pathway to Ascension*.

HANDLING "POSTULATES"

During our time in *the Physical Universe*, many of our *postulates* are automatically "stopped" by a simultaneously made *counter-postulate*. There are *reality-agreements* and *spiritual machinery* that contributes to this. But, this is what can actually keep our *energies* in "suspension" as a *wave-ridge* or *restriction* in the *flow*. In essence, it becomes like two of our *flows* colliding together, creating a *wave-peak* (like two land masses pushing together to make a mountain ridge).

What we are describing is really the *energetic-mass* that entangles our *Awareness* in *fragmentation* progressively throughout our *spiritual history*. It puts more of our *postulates*, our *Alpha-Thought*, our *spiritual power*, in a suspended state of *uncertainty*; our decisions to "*Be*" or "*Not-Be*" turn into "*maybe*." In other cases: we "*must have*" but "*can't have*" and now we "*don't have*" and then some of our *attention* remains perpetually fixed thereafter in a state of *confusion*.

• Let's *knowingly* practice a simple exercise that illustrates just some of what is taking place *unknowingly*. For this you will need a small object such as a *pencil*. We want to use a simple *postulate*, such as: "*reaching over and moving the pencil*."

–First: you will actually do it a few times; make the *postulate*, then move the *pencil*. Now: *postulate* "moving" it; then immediately change your mind, and *postulate* "not moving" it (and leave the *pencil* where it is).

–Next, we want to focus on mentally "relaxing" two *postulates* that are being "held" (or *maintained, created, &tc.*) in opposition to each other. Now: *postulate* both *intentions* simultaneously ("moving" and "not moving" the *pencil*) and "hold" them for a moment; then "relax."

As you hold the two *postulates* simultaneously: you may notice, or get the sense of, a *tension, mass,* or *solidity* forming that seems to dissipate or fall away when you relax. Part of this practice is learning how to *relax* the hold—or *release* the "charge"—on *postulates* that are held in suspension like this.

*** In *systematic processing*: *postulates* are typically handled like "*command-items*" of an *Implant* (see *Lesson-11*); and for good reason —"*command-items*" are basically *implanted postulates*. One of the reasons we use alternating *processing command-lines* in our *defragmentation* is to take *charge* off unseen *considerations* and *counter-intentions* too.

If there is still a doubt in the *Seeker's* thoughts as to their participation in the experience of *solidity* in *this Physical Universe*: *running* the following *postulates* as a *process* (each by themselves, alternating, then simultaneously and relaxing) probably won't make the world go away; but at least you'll have a better idea of what you're dealing with, and why you sometimes get a strong feeling that things aren't real when they are. [And if there was some obscure concise "*upper-level secret*" for this course, here in our final lesson, than this is it.]

> The most basic *postulate-counter-postulate* we all carry with us (*compulsively create*) that provides *Beta-Existence* its continuing *wave-ridge solidity* is:
>
>   ~*There is nothing there.*
>
>   ~*There is something there.*

POSITIVE "AFFIRMATIONS"

Perhaps the most appropriate parallel to *Alpha-Thought* that we find in the contemporary *New Age* and *New Thought* movements, is the emphasis on *"affirmations," "intention,"* and *"attraction."* These areas are not always handled or communicated properly; but this provides at least a basis for what we are talking about.

Repetitive *"positive affirmations"* are a way of applying *intention* over and over again, like hammering a nail, with a gentle *demand* that it become *reality.* This is very much like making a *postulate*; but in this case, a proper systematic handling of *"affirmations"* and *"positive intention"* is the most communicable and accessible way to treat the subject of *Alpha-Thought* as we bring this *Professional Course* to a close.

Any kind of *"charged" consideration* can affect your own view of yourself, or how you experience your own *reality* (or *Personal Universe*); and naturally, this alone can contribute to how you carry yourself in the *Game of Life*—what you do, how you act, *&tc.* This is what underlies all *"motivational coaching"* and similar things. But when most *Seekers* think of *Alpha-Thought*, it concerns a *spiritual ability* to directly influence *reality* itself.

Just as we have our own *spiritual machinery* (described in *Lesson-14*), there is larger-scale *spiritual machinery* for *manifesting reality.* If our personal *Implant-Platforms* are any indicator, this *reality-machinery* is likely operating on many *lines of programmed-code* (like *"command-items"* of an *Implant*) as an *"operating-system."* A lot of *mystics* have called it a *"universal mind"* or *"cosmic consciousness."* It may even be a fragmented part of ourselves that we threw into the "pool" when this *Beta-Existence* was constructed.

Most importantly: this *reality-machine* seems to arbitrarily project whatever is *imprinted* on it (or *repeated*) the most: taking what *persists* on the path that least *resists* toward manifestation. The idea behind *"positive affirmation"* is that if you repeat the *thought* enough and keep uploading into this *cosmic mind*, then you eventually shift the odds of the *reality probability-wave* a bit more in your favor. This is as much as can be reasonably expected given how many individuals in this *Shared-Universe* are also projecting *thoughts* and *intentions* into this *machinery*.

When handling any kind of *Alpha-Thought*, an individual has to be careful of their *exact wording*. Mistakes discovered early on with this likely led to an almost instinctual superstition surrounding *mispronouncing* or *misstating* any *"magic words."* Because the only time this is a real concern is when we compose the *intention* or apply *Alpha-Thought*; it certainly has nothing to do with some inability to pronounce foreign languages.

Just like a *postulate*, an effective *positive affirmation*—or *"Alphamation"*—must be stated as happening in present-time, not in the *future*. Saying *"I will be..."* is a *postulate* made for a *future* that will never manifest as the *present*. The *postulate* will go on perpetually as *"for the future."* *Reality* only consist of *"Now"* at any given moment—and so this very literal *reality-machine* really has no conception of anything related to "time." There is, however, usually some delay before anything manifests as *reality*.

Alpha-Thought is most effective when applied in favor of yourself or someone else; and never against someone. If you are in a *competition* or *game-condition*, you *affirm* or *postulate* in favor of yourself and your own ability; not that another's will be lesser, *&tc*. There are a lot of *"protections"* and *"defense mechanisms"* built into your own *Mind-System* and that of others, so it's best to improve yourself and not get involved with using *Alpha-Thought* for revenge or harm. That's one reason we ended up *down here* in the first place.

In most cases, *positive affirmations* and *Alpha-Thought* for one's own favor do not generally stir up trouble—except in cases of *fragmentation* in a the area you are treating. In extreme cases, a *positive affirmation* can trigger *fragmentation* attached to the *"negative"* side, and the *opposite* of what's *intended* will manifest. This is when even the most well-intended *New Ager* can still find themselves facing a *"psychic backlash"* from their efforts. If a *Seeker* finds this happening at all, they need to *spot* and *confront* the source of *fragmentation* before adding more *intention* (or *charge*) to that area. Use the appropriate *processing* as learned on the course.

When *"positive affirmations"* are made about one's *Self*, the energy cycles through our own *personal energy-system* and whatever *spiritual machinery* we maintain. It can *restimulate turbulence* from *unknown* and *unconfronted* areas and manifest "negative" results. A *"backlash"* is not just a "null" result where nothing happens; at the very least, it produces undesirable personal *"emotional states."*

For example: the *affirmation* "I am rich" might restimulate *charged considerations* that "only criminals are rich" or "rich people are all corrupt," *&tc.* that originate from this lifetime or earlier ones. And it is likely part of a *"chain of incidents"* not yet *confronted*, which leads to lingering *considerations* that are *unknowingly* maintained as ones' own *beliefs* or *postulates*. Perhaps you "embezzled from a company" in a former life, and "protested about wages" in another, then became a "loan shark" in one; this could all establish heavy *fragmentation* around the basic *postulate* of *"being rich."*

One way around our own *spiritual machinery* is to start off by making the *positive affirmation* for others; *intending* them to receive, or be endowed with, the characteristic or thing we want to manifest. This can be done with a specific person, or you can go to a public place and just make the *affirmation* a few times for each person you see. Since they are external to you, there is less chance of triggering your own *machinery*; and since it is a *positive affirmation* for another, it is not likely to trigger any of their *defense-mechanisms*.

There is another reason why you might want to start off with projecting a *"repetitive affirmation"* to an external target. Much like using a *"repetitive command-line"* in *systematic processing*, this focus of *attention* has an ability to bring other *associated data* and *consid-*

erations up to the "surface" for inspection/analysis. This means *postulating others "being rich"* may *resurface* the *fragmented consider-ations* (such as "all rich people are criminals") that need to be *confronted "As-It-Is"* in order to no longer remain a personal "*barri-er."*

Additionally, we can also approach our *goals* from different direc-tions. For example: if *"being rich"* is not working well as a *postulate/affirmation* directly, then perhaps *"being a more valuable employee"* (or something similar) might *"stick"*—and you still reach the same ends. However, it requires some work on your part still. Although *affirmations* may boost *confidence* and radiate an *"aura"* of *success* (that others can pick up on), it does not substi-tute having the skills or learning that may be necessary for applying it to *this Game*. So, an individual requires *Self-Honesty* and a *systematic* approach in order to effectively shape *reality*.

HANDLING THE "NEGATIVES"

We carry a lot of internal *"protections"* or *"buffers"* that prevent "idle" and "freewheeling" *thoughts* from instantly and spontan-eously manifesting within *Beta-Existence*. A single *thought* is generally not charged or powerful enough to be harmful or de-structive. We are permitted to *consider* things or *create mental imagery* in order to look at, or contemplate, without repercussions. Of course, if it *restimulates* some *fragmentation*, then that requires handling anyways. But it is a bad practice to go around putting a lot of effort into *avoiding* (or *withdrawing* from) a certain *thought* (or area) in fear that it will manifest negatively in *reality*.

While a single *thought* is not particularly destructive: *repetition* and *emotional charge* can certainly be. What we mean here is any *negat-ive postulates* that an individual likely *repeats* throughout everyday life—whether to themselves or in communication with others. For example: "I'm stupid" or "I'm always so clumsy"—these types of *postulates,* if *repeated* enough with conviction (and not as repetitive technique for *knowingly processing-out "charge"*), have a tendency to manifest (or *"stick"*) more easily in our experience of life.

Our *systematic processing* methods do treat a lot of "negative" expressions in our lives; generally, *fragmented* content is not particularly "fun" to handle. But when we *knowingly* treat things in *processing,* we also *systematically* balance it with the "positives." But both sides must be handled. This is because it is quite difficult to "push" the "positive" side of an area into reality when so much of the "negative" side has been reinforced and validated. Without treating the *charge* on an earlier *postulate* of *"being stupid,"* no gains can be expected from a single moment of now desiring to *"be smart."* Only when you can take all of the *charge* or *fragmentation* off of both sides of a former *consideration* will a *Seeker* be completely free to *postulate* anew.

We apply *"conceptual processing"* in *Traditional Piloting* to "loosen up" a certain area by *getting a sense* of the "concept" of actually "being" two different opposing conditions. For example: alternating *"get a sense of being rich"* and *"get a sense of being poor."* Alternating them this way in *processing* does not actually manifest these states or even add any *charge* to them. There is another reason we do this as well. In order to *postulate* a state of *"being rich,"* the *Alpha-Spirit* must necessarily *postulate* that a state of *"being poor"* also *exists*. This is the part that tends to get overlooked in *New Age* or *New Thought* materials; and the only remedy that we have found for effectively handling the "total package" of the *Human Condition* is the regimen of *systematic processing* described throughout the lessons of this course.

THE RITE OF ALPHA-THOUGHT

As we bring both this lesson and the entire *Professional Course* to a close, it seems like there should be "something" presented here at the end—that it has all been leading up to something "definitive" or "finite." Of course, the personal gains, *realizations*, and rehabilitation of *ability* takes place all throughout the course. And as has been suggested, a *Seeker* reaches a new level on their second pass through the course, starting from the beginning. It is not the *same* course when you do it again. Suddenly things start "clicking" better that seemed difficult before; or new deeper layers of *fragmentation* will *resurface*.

At the same time, we *are* reaching a completion-point of the actual instructional material given for this course. In other more *esoteric* traditions, a *Seeker* might receive an "initiation" ceremony, or get "installed" to a higher "grade" or "office" within the organization. Or, as a sign of "achievement," those traditions of a more *mystical* style might entrust a special *"spell-scroll"* or *"secret ritual"* that would somehow demonstrate a practical summation of all they had learned. This last one sounds like more *fun* for our purposes.

As a final technique, the present author has decided to incorporate a unique blend of *mystical* and *New Thought* methods to propose a "ritual" of sorts to aid a *Seeker* in more easily accessing *Alpha-Thought* and applying it to achieving their *goals*. It may be that applying what techniques still remain effective from our time in the *Magic Universe*, in order to play the *game* in *this Universe*, is the *only* way to even approach a "win" without mistakenly applying *force* against *force*, or becoming more ingrained in *reality-agreements* of *this world*. That is a matter we leave for a *Seeker* to determine as they achieve *higher levels* of *Knowing*.

By *"ritual"* or *"rite"* we mean a predetermined systematized arrangement of steps—not altogether different from the structure of a *"Formal Session."* There is no real *"occult"* tone intended otherwise. The only spirit summoned by this is the *Alpha-Spirit*, the highest level of *presence* of one's own *Self*, preferably as an *Actualized Awareness* that is free from heavy *fragmentation*.

This technique is no more "supernatural" than any of the others that we have discussed in prior lessons. However, it reflects a very ancient formula that has been held in veneration for its effectiveness. It is also not an inherent part of the *Pathway*, or any kind of *defragmentation* procedure. It is something one *does* as a part of participating in the *game* apart from being *in-session*. To the "uninitiated" that has not worked along the *Pathway* to this point, it may appear indistinguishable from other popular *visualization* techniques. It's power comes from the individual—from direct *Alpha-Thought*—and not by performing complicated actions or memorizing bad poetry to recite. Unsurprisingly, we will formalize this *rite* as *seven steps*.

STEP 1: *Clearly Define the Target and Postulate.*

Avoid "specific targets" (a *certain person* or a *particular company*). Use "general targets" (a *"girlfriend"* or a *"well paying job"*). Avoid referencing "time" in *postulates*—"having a job"; *not* "getting a job"—so that the *intention* and later *visualization* is of *"Now."* Avoid any "negative" *intentions* (such as against others).

STEP 2: *Defragment the Negative Side.*

Start with *conceptually running* alternate opposing statements. As good practice for more advanced work: a *systematic processing command-line* should negate the target or *terminal* and not the *condition*. For example: use *"I have a girlfriend / I have no girlfriend"* rather than *"I have a girlfriend / I don't have a girlfriend,"* because to be *systematically* correct, the "negative side" is also a *condition* of *"havingness."* It is also the "side" that an individual might still be strongly *postulating* for themselves. Any *turbulence* or *fragmentat-*

ion should be at least quieted down before proceeding. Use material from the course lessons to get the area under *control*.

STEP 3: *Generate Enthusiasm.*

In other traditions, this might be stated as *"raising personal energy."* We worked with *emotional states* and *Beta-Awareness* very early on this course. Naturally, we don't want to develop an obsessive desire or compulsion—but we have to *actually want* the *postulate* to come about if using a "positive" technique. When we expect, believe, or know something is happening, we tend to be more genuinely excited about it.

STEP 4: *See That Source Can Do It.*

Realizing that something is possible is the first step to making it *actual.* We are, each of us, *fragments* of an *Infinite Source,* balancing the *Nothingness* with the experience of an existence that contains potential *Everythingness.* If it helps a *Seeker* to *consider* that *"God"* can do it (as an expression of *Infinity*); or that the *"Cosmic Mind"* (*Reality-Machinery*) can produce it, just go with what's comfortable. Whatever it is you *conceive* as *manifesting reality,* know that what you are *postulating* is within their capabilities to manifest.

STEP 5: *See That You Are Source.*

Postulates are *causative thought*; and they only *"stick"* when projected strongly from a *Cause* or "source-point." Whatever it is you *conceive* as *"Source"*—and even *"God"* for this *Universe,* or *"Reality-Making Machinery"*—know that *you* are a part of *IT,* an integral interactive part of whatever *"IT"* is. There are a lot of other individuals that are a part of it too. But the *"drops"* you add to this *"pool"* have the ability to permeate the entire *"ocean."* Your *postulates* have an ability to ripple across the *Spheres of Existence* into real manifestation.

STEP 6: *See It As Happening.*

Now, *you,* as *Source,* then *"see"* that the thing or target-item is "*done*" or "*complete*"—*visualizing* or *creating mental imagery* of it *happening now* (not at some future point). Consider the example with the "door" earlier in this lesson: we aren't *postulating* that it will eventually happen; but that it is happening (or has happened right in front of us). In *Beta-Existence*, you *postulate/demand* it *exist* within that "*reality-wave*" that is constantly being generated. It might take some time, given the manifestation delays of *this reality*, but the *intention* is made and complete.

STEP 7: *Acknowledgment.*

The final step of this ritual is the completion of a cycle: we express *gratitude* and *acknowledgment* as though it has been materially presented to us and received. Recall the last time you received something you were expecting or wanting very much. That is the concept we want to project before ending our ritual. Whether we see it as being directed to "*God*" or a "*group mind*" of some sort, we *acknowledge* that it has happened; that we are *grateful* for our manifestations brightening not only our own lives, but also enriching the *Game* that we all share. Then take the culmination of your whole ritual—almost as though it were an *implanting-incident*—and strongly *project* or "*upload*" it into the *Universal Reality* as a single *postulate*, and completely "release it" (or "let go") knowing it is done.

This lesson marks the completion of
The New Standard Systemology Professional Course.

You are invited to continue the upper levels with
The "Keys to the Kingdom" Advanced Training Course.

SYSTEMOLOGY
TECHNICAL
DICTIONARY

This complete technical dictionary
may be used while studying
this workbook and any
other Systemology materials.

THE COMPLETE SYSTEMOLOGY
TECHNICAL DICTIONARY
(VERSION 5.0)

—A—

A-for-A (one-to-one) : an expression meaning that what we say, write, represent, think or symbolize is a direct and perfect reflection or duplication of the actual aspect or thing—that "A" is for, means and is equivalent to "A" and not "a" or "q" or "!"; in the relay of communication, the message (or messenger particle) is sent and perfectly duplicated in form and meaning when received.

aberration : a departure from what is right or correct; a deviation from, or distortion in, what is true or right or straight; in chromatic light science, the failure of a mirror, lens or refracting surface to produce an exact "*one-to-one*" or "*A-for-A*" duplication between an object and its image; in *NexGen Systemology*, a term to describe *fragmentation* as it applies to an individual, which causes them to "stray" form the *Pathway* (and experience a *fragmented reality*).

abreaction (abreactive therapy) : the "burn off" or "purging" or "discharge" of "unconscious" (reactive response) as applied to early 20th century German psychology, from *abreagieren*, meaning "coming down" from a release or expression of a repressed or forgotten emotion; in *NexGen Systemology*, fully "resurfacing" traumatic past experiences consciously (on one's own determinism) in order to purge them of their emotional excess (or "charge"); also "*Route-1*" and "*catharsis.*"

acid-test : a metaphor refers to a chemical process of applying harsh nitric acid to a golden substance (sample) to determine its genuineness; in *NexGen Systemology*, an extreme conclusive process to determine the reality, genuineness or truth of a substance, material, particle or piece of information.

acknowledgment : a response-communication establishing that an immediately former communication was properly received, duplicated and understood; the formal acceptance and/or recognition of a communication or presence.

activating event : an incident or occurrence that automatically stimulates a conscious or unrecognized reminder or 'ping' from an earlier *imprinting incident* recorded on one's own personal timeline as an emotionally charged and encoded memory; an incident or instance when thought systems are activated to determine the consequence or significance of an activity, motion or event—often demonstrated as *Activating Event → Belief Systems → Consideration.*

actualization : to make actual, not just potential; to bring into full solid Reality; to realize fully in *Awareness* as a "thing."

affinity : the apparent and energetic *relationship* between substances or bodies; the degree of *attraction* or repulsion between things based on natural forces; the *similitude* of frequencies or waveforms; the degree of *interconnection* between systems.

agreement (reality) : unanimity of opinion of what is "thought" to be known; an accepted arrangement of how things are; things we consider as "real" or as an "is" of "reality"; a *consensus* of what is real as made by standard-issue (common) participants; what an individual contributes to or accepts as "real"; in *NexGen Systemology*, a synonym for "*reality.*"

allegorical : a representation of the abstract, metaphysical or "spiritual" using physical or concrete forms.

alpha : the first, primary, basic, superior or beginning of some form; in *NexGen Systemology*, referring to the state of existence operating on spiritual archetypes and postulates, will and intention "exterior" to the low-level condensation and solidarity of energy and matter as the 'physical universe'.

alpha control center (ACC) : the highest relay point of *Beingness* for an individuated *Alpha-Spirit*, *Self* or "I-AM"; in *NexGen Systemology*—a point of spiritual separation of ZU at (7.0) from the *Infinity of Nothingness* (8.0); the truest actualization of *Identity*; the highest *Self-directed* relay of *Alpha-Self* as an *Identity-Continuum*, operating in an *alpha-existence* (or "Spiritual Universe"–AN) to *determine* "Alpha Thought" (6.0) and WILL-*Intention* (5.0) *exterior* to the "Physical Universe"–(KI); the "wave-peak" of "I" emerging as individuated consciousness from *Infinity.*

alpha-spirit : a "spiritual" *Life*-form; the "true" *Self* or I-AM; the *individual*; the spiritual (*alpha*) *Self* that is animating the (*beta*) physical body or "*genetic vehicle*" using a continuous *Lifeline* of spiritual ("*ZU*") energy; an individual spiritual (*alpha*) entity possessing no physical mass or measurable waveform (motion) in the Physical Universe as itself, so it animates the (*beta*) physical body or "*genetic vehicle*" as a catalyst to experience *Self*-determined causality in effect within the *Physical Universe*; a singular unit or point of *Spiritual Awareness* that is *Aware* that it is *Aware*.

alpha thought : the highest spiritual *Self-determination* over creation and existence exercised by an Alpha-Spirit; the Alpha range of pure *Creative Ability* based on direct postulates and considerations of *Beingness*; spiritual qualities comparable to "thought" but originating in Alpha-existence (at "6.0") independently superior to a *beta-anchored* Mind-System, although an Alpha-Spirit may use Will ("5.0") to carry the intentions of a postulate or consideration ("6.0") to thc Master Control Center ("4.0").

amplitude : the quality of being *ample*; the size or amount of energy that is demonstrated in a *wave.* In the case of audio waves, we associate amplitude with "volume." It is not a statement about the frequencies of waves, only how "loud" they are—to what extent they are or may be projected (or audible).

AN : an ancient "Sumerian" cuneiform sign for Heaven or "God"; in *Mardukite Zuism and Systemology* designating the *'spiritual zone'* (or *'Alpha Existence'*); the *Spiritual Universe*—comprised of spiritual matter and spiritual energy; a direction of motion toward spiritual *Infinity*, away from or superior to the physical (*'KI'*); the spiritual condition of existence providing for our primary *Alpha* state as an individual *Identity* or *I-AM-Self* which interacts and experiences *Awareness* of a *beta* state in the *Physical Universe* (*'KI'*) as *Life*.

anathema : a thing or person to be detested, loathed or avoided; a thing or person accursed or despised such as to wish damnation or "divine punishment" upon.

anchor (conceptual) : a stable point in space; a fixed point used to hold or stabilize a spatial existence of other points; a spatial point that fixes the parameters of dimensional orientation, such as the corner-points of a solid object in relation to other points in space; in *NexGen Systemology*, "beta-anchored" is an expression used to describe the fixed orientation of a viewpoint from Self in relation to all possible spatial points in *beta-existence* ("physical universe"), or else the existential points that fix the operation of the "body" within the space-time of *beta-existence*.

Ancient Mystery School : the original arcane source of all esoteric knowledge on Earth, concentrated between the Middle East and modern-day Turkey and Transylvania c. 6000 B.C. and then dispersing south (Mesopotamia), west (Europe) and east (Asia) from that location.

antinomian : a term applied to *Gnostics* (popularized by Martin Luther during the Christian reformation) denoting a rejection of formal religious morals and dogma—decreed, written and interpreted by humanity—as a true pathway to Ascension (some elements appear in all forms of religious protest and reformation but as an extreme, would be considered spirto-religious rebellious punkdom by some modern standards, but it should be understood that it does follow a higher ethic, such as Mardukite Utilitarianism.

apotheosis : from the *Greek* word, meaning *"to deify"*; the highest point or apex (for example, of "true knowledge" and "true experience"); an ultimate development of; a glorified or "deified" *ideal*, such as is a quality of *godhood*.

apparent : visibly exposed to sight; evident rather than actual, as presumed by Observation; readily perceived, especially by the senses.

a-priori : from "cause" to "effect"; from a general application to a particular instance; existing in the mind prior to, and independent of experience or observation; validity based on consideration and deduction rather than experience.

archetype : a "first form" or ideal conceptual model of some aspect; the ultimate prototype of a form on which all other conceptions are based.

ascension : actualized *Awareness* elevated to the point of true "spiritual existence" exterior to *beta existence*. An "Ascended Master" is one who has returned to an incarnation on Earth as an inherently *Enlightened One*, demonstrable in their actions—they have the ability to *Self-direct* the "Spirit" as *Self*, just as we are treating the "Mind" and "Body" at this current grade of instruction; previously treated in *Moroii ad Vitam* as a state of Beingness after *First Death*, experienced by an *etheric body*, which is able to maintain consciousness as a personal identity continuum with the same *Self-directed* control and communication of Will-Intention that is exercised, actualized and developed deliberately during one's present incarnation.

assessment : an analysis or synthesis of collected information, usually about a person or group, in relation to an *assessment scale*.

assessment scale : an official assignment of graded/gradient numeric values correlated to specific tiers with individual preassigned meanings.

associative knowledge : significance or meaning of a facet or aspect assigned to (or considered to have) a direct relationship with another facet; to connect or relate ideas or facets of existence with one another; a reactive-response image, emotion or conception that is suggested by (or directly accompanies) something other than itself; in traditional systems logic, an equivalency of significance or meaning between facets or sets that are grouped together, such as in *(a + b) + c = a + (b + c)*; in NexGen Systemology, erroneous associative knowledge is assignment of the same value to all facets or parts considered as related (even when they are not actually so), such as in *a = a, b = a, c = a* and so forth without distinction.

assumption : the act of taking or gathering to one's Self; taking possession of, receive or behold.

attenergy : *NexGen Systemological NewSpeak* for "attention energies"; the flow of consciousness "energy" that is directed as "attention"; semantic recognition of an axiom from the *Arcane Tablets* that states: "energy flows where attention goes."

attention : active use of *Awareness* toward a specific aspect or thing; the act of "attending" with the presence of *Self*; a direction of focus or concentration of *Awareness* along a particular channel or conduit or toward a particular terminal node or communication termination point; the Self-directed concentration of personal energy as a combination of observation, thought-waves and consideration; focused application of *Self-Directed Awareness*.

authoritarian : knowledge as truth, boundaries and freedoms dictated to an individual by a perceived, regulated or enforced "authority."

auto-suggestion (self-hypnosis) : auto-conditioning; self-programming; delivering directed affirmations or statements repeatedly to *Self* in order to condition a change in behavior or beliefs; any *Self-directed* technique intended to generate a specific "*post-hypnotic suggestion.*"

awareness : the highest sense of-and-as Self in knowing and being as I-AM (the *Alpha-Spirit*); the extent of beingness directed as a POV experienced by Self as knowingness.

axiom : a fundamental truism of a knowledge system, esp. *logic*; all *maxims* are also *axioms*; knowledge statements that require no proof because their truth is self-evident; an established law or systematic principle used as a *premise* on which to base greater conclusions of truth.

—B—

Babylonian : the ancient Mesopotamian civilization that evolved from *Sumer*; inception point for systematization of civic society and religion.

Back-Scan : to apply Awareness, *Zu-Vision* or "Alpha-Sight" (*exterior* to the *Human Condition*) and *resurface* impressions for recreating *Mental Imagery* of the *Backtrack* within one's own Personal Universe and treat with Wizard-Level (*Grade-V+*) methodology.

Backtrack : to retrace one's steps or go back to an early point in a sequence; an applied spiritual philosophy within *Metahuman Systemology "Wizard Grades"* regarding continuous existence of an individual's "*Spiritual Timeline*" through all lifetime-incarnations; the course that is already laid behind us; a methodology of systematic processing methods developed to assist in revealing "hidden" *Mental Images* and *Imprints* from one's past and reclaim attention-energies "left behind" with them by increasing ability to manage and control personal energy mechanisms fixed to their continuous automated creation.

band : a division or group; in *NexGen Systemology*, a division or set of frequencies on the ZU-line that are tuned closely together and referred to as a group.

BAT (Beta-Awareness Test) : a method of *psychometric evaluation* developed for *Mardukite Systemology* to determine a "basic" or "average" state of personal *beta-Awareness*; first developed for the text "*Crystal Clear.*"

"bell, book & candle" : three dissimilar objects that are kept accessible during a processing session (the book is often a copy of *The Systemology Handbook* or a hardcover copy of *The Tablets of Destiny* with the dust-jacket removed if it is less distracting that way); a term meant to indicate a Pilot's "objective processing kit" of objects generally present in the session room (accessible on a shelf, table or pedestal stands); in *NexGen Systemology,* the name of an objective processing philosophy pertaining to command of personal reality; historically, a formal ritual used by the Roman Catholic church to ceremonially declare an individual "guilty of the most heinous sins" as "excommunicated (to hold no further communications with) by anathema"—whereby a *bell* is rung, a *holy book* is closed and all *candles* are snuffed out—thus we therapeutically use the same symbolism historically representing religious fragmentation for modern systematic defragmentation purposes.

beta (awareness) : all consciousness activity ("*Awareness*") in the "Physical Universe" (KI) or else *beta-existence*; *Awareness* within the range of the *genetic-body*, including material thoughts, emotional responses and physical motors; personal *Awareness* of physical energy and physical matter moving through physical space and experienced as "time"; the *Awareness* held by *Self* that is restricted to a physical organic *Lifeform* or "*genetic vehicle*" in which it experiences causality in the *Physical Universe*.

beta (existence) : all manifestation in the "Physical Universe" (KI); the "Physical" state of existence consisting of vibrations of physical energy and physical matter moving through physical space and experienced as "time"; the conditions of *Awareness* for the *Alpha-spirit* (*Self*) as a physical organic *Lifeform* or "*genetic vehicle*" in which it experiences causality in the *Physical Universe*.

beta-defragmentation : toward a state of *Self-Honesty* in regards to handling experience of the "Physical Universe" (*beta-existence*); an applied spiritual philosophy (or technology) of Self-Actualization originally described in the text "*Crystal Clear*" (*Liber-2B*), building upon theories from "*Systemology: The Original Thesis.*"

biological unconsciousness : the organism independent of the sentient *Awareness* of the *Self* to direct it; states induced by severe injury and anesthesia.

biomagnetic/biofeedback : a measurable effect, such as a change in electrical resistance, that is produced by thoughts, emotions and physical behaviors which generate specific 'neurotransmitters' and biochemical reactions in the brain, body and across the skin surface.

—C—

cacophony : dissonant, turbulent, harsh and/or discordant sound or noise.

calcified : in nature, to calcify is to harden like stone from calcium and lime deposits; in philosophic applications, refers to a state of hardened fixed bone-like inflexibility; a condition change to rigidly solid.

capable : the actual capacity for potential ability.

CAT / "Creative Ability Test" : a method of increasing personal freedom and unlimited creative potential of the Alpha-Spirit (Self) independent and exterior to conditions and reality agreements with beta-existence; a Wizard-Level training regimen first developed for the Grade-IV text "*Imaginomicon*" (*Liber-3D*).

catalog / catalogue : a systematic list of knowledge or record of data.

catalyst : something that causes action between two systems or aspects, but which itself is unaffected as a variable of this energy communication; a medium or intermediary channel.

catharsis / cathartic processing : from the Greek root meaning "pure" or "perfect"; Gnostic practices of "consolamentum" where an individual removes distorting/fragmented emotional charges and encoding from a personal energy flow/circuit connected or associated with some terminal, mass, thing, *&tc.*; in *NexGen Systemology*, the emptying out or discharge of emotional stores; also "*abreaction*" or "*Route-1.*"

causative : as being the cause; to be at cause.

chakra : an archaic Sanskrit term for "wheel" or "spinning circle" used in *Eastern* wisdom traditions, spiritual systems and mysticism; a concept retained in NexGen Systemology to indicate etheric concentrations of energy into wheel-mechanisms that process *ZU* energy at specific frequencies along the *ZU-line*, of which the *Human Condition* is reportedly attached *seven* at various degrees as connected to the Gate symbolism.

channel : a specific stream, course, current, direction or route; to form or cut a groove or ridge or otherwise guide along a specific course; a direct path; an artificial aqueduct created to connect two water bodies or water or make travel possible.

charge : to fill or furnish with a quality; to supply with energy; to lay a command upon; in *NexGen Systemology*—to imbue with intention; to overspread with emotion; application of *Self-directed (WILL)* "intention" toward an emotional manifestation in beta-existence; personal energy stores and significances entwined as fragmentation in mental images, reactive-response encoding and intellectual (and/or) programmed beliefs; in traditional mysticism, to intentionally fix an energetic resonance to meet some degree, or to bring a specific concentration of energy that is transferred to a focal point, such as an object or space.

circuit : a circular path or loop; a closed-path within a system that allows a flow; a pattern or action or wave movement that follows a specific route or potential path only; in *NexGen Systemology*, "*communication processing*" pertaining to a specific flow of energy or information along a channel; *see* also "*feedback loop.*"

Circuit-1 : in *Grade-IV* "communication processing" (introduced in *Metahuman Destinations* as *Route-3*), the flow of energy and information connected to outflow, what *Self* has expressed, projected outwardly or done.

Circuit-2 : in *Grade-IV* "communication processing" (introduced in *Metahuman Destinations* as *Route-3*), the flow of energy and information connected to inflow, what "others" have done to *Self,* what it has received inwardly or had *happen to.*

Circuit-3 : in *Grade-IV* "communication processing" (introduced in *Metahuman Destinations* as *Route-3*), the flow of energy and information connected to cross-flows, what *Self* has witnessed of others (or another) projecting or doing toward others (or another).

Circuit-0 : a more advanced concept introduced to *Grade-IV* "communication processing" (as listed on SOP-2C in *Metahuman Destinations* for "*Pre-A.T*" or "*Route-0*" applications), which targets *'postulates'* and *'considerations'* generated and stored by *Self* for *Self* and the direction, energy or flows representing what *Self* "does" for and/or to *Self.* This circuit is treated further in *Wizard Level* work,

chronologically : concerning or pertaining to "time"; to treat as "units" of "time" ; to sequence a series of events or information with regard to the order it happened or originated (in time).

clockwork : rigidly fixed gear-like systems that operate mechanically and directly upon one another to function; a "clockwork universe theory" is a "closed-system design" popular in Newtonian Physics attributes all actions of energy-matter in space-time as reactions in accordance with a "Divine Decree" or fixed design that functions like a "clock-mechanism" and does not account for the "Observer."

code (ethics) : an outline of *ethical* standards regarding social participation and acceptable behavior; not generally enforced as *law* itself, but a standard that reasonable individuals are actualized (or civil) enough to *Self-Determine* (by choice) their own following (or adherence) if it is *right* and *good*; shared reality agreements that promote optimum conditions of continued existence ("SURVIVAL" in *Beta-existence*; "CREATION" in *Alpha*) for the highest affected "Sphere of Existence" (on the *Standard Model*).

codification : process of collecting, analyzing and then arranging knowledge in a standardized and more accessible systematic form, often by subject, theme or some other designation.

collapsing a wave : also, *"wave-function collapse"*; in *Quantum Physics*, the concept that an Observer is "collapsing" the wave-function to something "definite" by measuring it; defining or calcu-lating a wave-function or interaction of potential interactions by an Observation; in *NexGen Systemology*, when a wave of potentiality or possibility because a finite fixed form; Consciousness or *Aware-ness* "collapses" a wave-function of energy-matter as a necessary "third" Principle of Apparent Manifestation (first described in *"Tablets of Destiny"*); potentiality as a wave is collapsed into an ap-parent *"is"*, the energy of which is freed up in systematic processing by *"flattening"* a "collapsed" wave back into its state of potentiality.

command : in *Metahuman Systemology*, responsibility and ability of Self (I-AM) as operating from its ideal "exterior" *Point-of-View* as Alpha Spirit; to direct communication for control of the *genetic vehicle* and Mind-Body connection that is perfectly duplicated from a source-point to a receipt-point along the ZU-line.

command line : see *"processing command line"* (PCL).

common knowledge (game theory) : facts that all "players" know, and they know that all other "players" also know—such as the very structure of the "game" being played.

communication : successful transmission of information, data, en-ergy (&tc.) along a message line, with a reception of feedback; an energetic flow of intention to cause an effect (or duplication) at a distance; the personal energy moved or acted upon by will or else 'selective directed attention'; the 'messenger action' used to transmit and receive energy across a medium; also relay of energy, a mes-sage or signal—or even locating a personal POV (viewpoint) for the Self—along the *ZU-line*.

communication (circuit) processing : a methodology of Grade-IV Metahuman Systemology that emphasizes analysis of all Mind-Sys-tem energy flows (information) transmitted and stored along cir-cuits of a channel toward some terminal, thing or concept, particularly: what Self has out-flowed, what Self has in-flowed, and the cross-flows that Self has observed; also *"Route-3"*

compulsion : a failure to be responsible for the dynamics of control —starting, stopping or altering—on a particular channel of communication and/or regarding a particular terminal in existence; an energetic flow with the appearance of being 'stuck' on the action it is already doing or by the control of some automatic mechanism.

computing device : a calculator or modern computer; a mechanism that performs specific functions, particularly input, output and storage of data/information.

concept : a high-frequency thought-wave representing an "idea" which persists because it is not restricted to a unique space-time; an abstract or tangible "idea" formed in the "Mind" or *imagined* as a means of understanding, usually including associated "Mental Images"; a seemingly timeless collective thought-theme (or subject) that entangles together facets of many events or incidents, not just a single significant one.

conceptual processing : a Wizard-Level methodology introduced intermittently throughout materials of Metahuman Systemology that emphasizes fully "getting the sense of" (or "contacting the idea of") a particular condition as prompted by a PCL and on one's own determination; a systematic practice-drill regarding considerations and postulates (Alpha Thought) regarding various reality agreements; a *Route-0* variant employing *Creativeness* and *Imagination* for systematic processing; also *Route-0E* when used for *Ethics Processing.*

condense (condensation) : the transition of vapor to liquid; denoting a change in state to a more substantial or solid condition; leading to a more compact or solid form.

condition : an apparent or existing state; circumstances, situations and variable dynamics affecting the order and function of a system; a series of interconnected requirements, barriers and allowances that must be met; in "contemporary language," bringing a thing toward a specific, desired or intentional new state (such as in "conditioning"), though to minimize confusion about the word "condition" in our literature, *NexGen Systemology* treats "contemporary conditioning" concepts as imprinting, encoding and programming.

conflict : the opposition of two forces of similar magnitude along the same channel or competing for the same terminal; the inability to duplicate another POV; a thought, intention or communication that is met with an opposing counter-thought or counter-intention that generates an energetic cluster.

confront : to come around in front of; to be in the presence of; to stand in front of, or in the face of; to meet "face-to-face" or "face-up-to"; additionally, in *NexGen Systemology*, to fully tolerate or acceptably withstand an encounter with a particular manifestation or encounter.

consciousness : the energetic flow of *Awareness*; the Principle System of *Awareness* that is spiritual in nature, which demonstrates potential interaction with all degrees of the Physical Universe; the *Beingness* component of our existence in *Spirit*; the Principle System of *Awareness* as *Spirit* that directs action in the Mind-System.

consensual (consensus) : formed or existing simply by consent— by general or mutual agreement; permitted, approved or agreed upon by majority of opinion; knowingly agreed upon unanimously by all concerned; to be in agreement on the objective universe and/or a course of action therein.

consideration : careful analytical reflection of all aspects; deliberation; determining the significance of a "thing" in relation to similarity or dissimilarity to other "things"; evaluation of facts and importance of certain facts; thorough examination of all aspects related to, or important for, making a decision; the analysis of consequences and estimation of significance when making decisions; in *NexGen Systemology*, the postulate or Alpha-Thought that defines the state of beingness for what something "*is.*"

continuity : being a continuous whole; a complete whole or "total round of"; the balance of the equation ["–120" + "120" = "0" &tc.]; an apparent unbroken interconnected coherent whole; also, as applied to Universes in *NexGen Systemology*, the lowest base consideration of space-time or commonly shared level of energy-matter apparent in an existence, or else the lowest degree of solidity or condensation whereby all mass that exists is identifiable or communicable with all other mass that exists; represented as "0" on the *Standard Model* for the Physical Universe (*beta-existence*), a level of existence that is below Human emotion, comparable to the solidity of "rocks" and "walls" and "inert bodies."

continuum : a continuous enduring uninterrupted sequence or condition; observing all gradients on a *spectrum*; measuring quantitative variation with gradual transition on a spectrum without demonstrating discontinuity or separate parts.

control (general) : the ability to start, change or start some action or flow of energy; the capacity to originate, change or stop some mode of human behavior by some implication, physical or psychological means to ensure compliance (voluntarily or involuntarily).

control (systems) : communication relayed from an operative center or organizational cluster, which incites new activity elsewhere in a system (or along the *ZU-line*).

correlate : a relationship between two or more aspects, parts or systems.

correspondence : a direct relationship or correlation; see also "*associative knowledge.*"

Cosmic History : the entire continuous *Spiritual Timeline* of all existence, starting with the *Infinity of Nothingness* and individuation of Self and its Home Universe, running through various Games Universes and ultimately leading to condensation and solidification of this Physical Universe experienced in present-time.

Cosmic Law : the "Law" of Nature (or the Physical Universe); the "Law" governing cosmic ordering; often called "Natural Law" in sciences and philosophies that attempt to codify or systematize it.

cosmology : a systematic philosophy defining origins and structure of an apparent Universe.

Cosmos : archaic term for the "Physical Universe"; semantically implies chaos brought into order; in *NexGen Systemology*, can also include considerations of "Universes" experienced previously as a *beta-existence*.

counter-productive : contrary to the greater or original purpose or intention; in *NexGen Systemology*, anything which brings *Life* away from its sustainable goal or position of *Infinite Existence*.

crash-coursed : a very intense or steep delivery of education over a very brief time period, usually applied to bring a student "up-to-speed" or "up-to-date" for receiving and understanding newer or cumulatively more advanced material.

creative ability test : see "*CAT.*"

creativeness processing : a *systematic processing* methodology introduced in *Grade-IV Metahuman Systemology* (*Wizard Level-0*) that emphasizes personal use of "*Imagination,*" or else "creative ability" of Self and freeing considerations of the Alpha-Spirit to *Be* or *Create* anything within its Personal Universe, independent of reality agreements with beta-existence; also "*Route-0.*"

Crossing the Abyss : to enter the spiritual or metaphysical unknown in "Self-annihilation" to purify the Self and "return to the Source."

Crystal Clear : the second professional publication of Mardukite Systemology, released publicly in December 2019; the second professional text in Grade-III Mardukite Systemology, released as "*Liber-2B*" and reissued in the Grade-III Master Edition "*Systemology Handbook*"; contains fundamental theory of "*Beta-Defragmentation*" and "*Route-2*" systematic processing methodology.

cuneiform : the oldest extant writing system at the inception of modern civilization in Mesopotamia; a system of wedge-shaped script inscribed on clay tablets with a reed pen, allowing advancements in record keeping and communication no longer restricted to more literal graphic representations or pictures.

cuneiform signs : the cuneiform script, as used in ancient Mesopotamia, is not represented in a linear alphabet of "letters," but by a systematic use of basic word "signs" that are combined to form more complex word "signs"—each sign represented a "sound" more than it did a letter, such as "ab," "ad", "ba", "da" &tc.

—D—

data-set : the total accumulation of knowledge used to base Reality.

dead-memories : outdated/inadequate/erroneous data.

defragmentation : the *reparation* of wholeness; collecting all dispersed parts to reform an original whole; a process of removing "*fragmentation*" in data or knowledge to provide a clear understanding; applying techniques and processes that promote a *holistic* interconnected *alpha* state, favoring observational *Awareness* of continuity in all spiritual and physical systems; in *NexGen Systemology*, a "*Seeker*" achieving an actualized state of basic "*Self-Honest Awareness*" is said to be *beta-defragmented*, whereas *Alpha-defragmentation* is the rehabilitation of the *creative ability*, managing the *Spiritual Timeline* and the POV of *Self* as Alpha-Spirit (I-AM); see also "*Beta-defragmentation*."

degree : a physical or conceptual *unit* (or point) defining the variation present relative to a *scale* above and below it; any stage or extent to which something *is* in relation to other possible positions within a *set* of "*parameters*"; a point within a specific range or spectrum; in *NexGen Systemology*, a *Seeker's* potential energy variations or fluctuations in thought, emotional reaction and physical perception are all treated as *"degrees."*

demographics : segments of the population uniquely identified, whether real or representative; targeting a specific portion of the population, such as for marketing or statistics.

destiny : what is set down, made firm, standard, or stands fixed as a constant end; the absolute *destination* regardless of whatever course is traveled; in *NexGen Systemology*, the *"destiny"* of the *"Human Spirit"* (or *"Alpha Spirit"*) is infinite existence—*"Immortality."*

dichotomy : a division into two parts, types or kinds.

differential : the quantitative value difference between two forces, motions, pressures or degrees.

differentiation : an apparent difference between aspects or concepts.

discernment : to perceive, distinguish and/or differentiate experience into true knowledge.

displace : to compel to leave; to move or replace something with something else in its place or space.

dissonance : discordance; out of step; out of phase; disharmonious; the "differential" between the way things are and the way things are experienced; cognitive dissonance could be demonstrated as A = abc, or C = A, the duplication of truth/communication is not A-for-A.

dogma : religious doctrines or opinion-based beliefs (data-set) treated socially as fact, especially regarding "divinity" or "God" (the common "Human" interpretation of the "domain" of Infinity) represented by the "Eighth Sphere" on our original Standard Model of Systemology; religiously defined values, taboos and ethical standards emphasized by cultural/religious socialization and mythographic beliefs (even above any observable causal effects, logical sequences or verifiable proofs).

dramatization / dramatize : a vivid display or performance as if rehearsed for a "play" (on stage); a *'circuit'* recording *'imprinted'* in the past and, once restimulated by a facet of the environment, the individual "replays" it as through reacting to it in the present (and identifying that reality as present reality); acts, actions and observable behaviors that demonstrate identification with a particular character type, "phase" or personality program; a motivated sequence-chain, implant series or imprinted cycle of actions—usually irrational or counter-survival—repeated by an individual as it had previously happened to them; a reoccurring or reactively triggered out-flow, communication or action that indicates an individual "occupying" a particular *'Point-of-View'* (*POV*)—typically fixed to a specific (past) identification (identity) that is space-time locatable (meaning a point where significant *Attenergy*—enough to compulsively create and maintain a POV—is "stuck" or "hung up" on the *BackTrack*).

dross : prime material; specifically waste-matter or refuse; the discarded remains collected together.

dynamic (systems) : a principle or fixed system which demonstrates its *'variations'* in activity (or output) only in constant relation to variables or fluctuation of interrelated systems; a standard principle, function, process or system that exhibits *'variations'* and change simultaneously with all connected systems; each *'Sphere of Existence'* is a dynamic system, systematically affecting (supporting) and affected (supported) by other *'Spheres'* (which are also dynamic systems).

—**E**—

Eastern traditions : the evolution of the *Ancient Mystery School* east of its origins, primarily the Asian continent, or what is archaically referred to as "oriental."

echelon : a level or rung on a ladder; a rank or level of command.

eclipse : to cast a shadow or darken; to block out or obscure a comparison.

EDA : "electro-dermal activity"; see also *GSR-Meter.*

electro-psychometer ("E-meter") : see *GSR-Meter.*

elocution : the skillful use of clearly directed and expressive speech; the expert demonstration of articulation, pronunciation and dictation to express a message.

emotional encoding : the readable substance/material (data) of *'imprints'*; associations of sensory experience with an *imprint*; perceptions of our environment that receive an *emotional charge*, which form or reinforce facets of an *imprint*; perceptions recorded and stored as an *imprint* within the "emotional range" of energetic manifestation; the formation of an energetic store or charge on a channel that fixes emotional responses as a mechanistic automation, which is carried on in an individual's *Spiritual Timeline* (or personal continuum of existence).

enact : to make happen; to bring into action; to make part of an act.

encompassing : to form a circle around, surround or envelop around.

end point : the moment when the goal of a process has been achieved and to continue on with it will be detrimental to the gains; the finality of a process when the *Seeker* has achieved their optim-um state from the current cycle (whether or not they run through it again at a later date with a different level of *Awareness* or know-ledge base doesn't change the fact that it has flattened the standing wave

energetic exchange : communicated transmission of energetically encoded "information" between fields, forces or source-points that share some degree of interconnectivity; the event of "waves" acting upon each other like a force, flowing in regard to their proximity, range, frequency and amplitude.

energy signatures : a distinctive pattern of energetic action.

enforcement : the act of compelling or putting (effort) into force; to compel or impose obedience by force; to impress strongly with ap-plications of stress to demand agreement or validation; the lowest-level of direct control by physical effort or threat of punishment; a low-level method of control in the absence of true communication.

engineering : the *Self-directed* actions and efforts to utilize know-ledge (observed causality/science), maths (calculations/quantifica-tion) and logic (axioms/formulas) to understand, design or manifest a solid structure, machine, mechanism, engine or system; as "*Real-ity Engineering*" in *NexGen Systemology*—intentional *Self-directed* adjustment of existing Reality conditions; the application of total *Self-determinism* in *Self-Honesty* to change apparent Reality using fundamentals of *Systemology* and *Cosmic Law*.

entanglement : tangled together; intertwined and enmeshed sys-tems; in *NexGen Systemology*, a reference to the interrelation of all particles as waves at a higher point of connectivity than is apparent, since wave-functions only "collapse" when someone is *Observing*, or doing the measuring, evaluating, &tc.

entropy : the reduction of organized physical systems back into chaos-continuity when their integrity is measured against space over time; reduction toward a zero-point.

epicenter : the point from which shock-waves travel.

epistemology : a school of philosophy focused on the truth of knowledge and knowledge of truth; theories regarding validity and truth inherent in any structure of knowledge and reason; the original "school of philosophy" from which all other "disciplines" were derived; the study of knowing how to know knowledge, reason and truth.

erroneous : inaccurate; incorrect; containing error.

esoteric : hidden; secret; knowledge understood by a select few.

etching : to cut, bite or corrode with acid to produce a pattern.

ethics : an intellectual philosophy concerning *rightness* and *wrongness* based on "logic" and "reason" (rationale) combined with observable consequences and tendencies of action or conduct; formal name for a "moral philosophy" (study of moral choices); in ancient times, originally treated *one-to-one* with "Cosmic Law" regarding *causation*, *order* and *sequence*; an objective (Universal) philosophy of *rightness* and *wrongness*, treated separate from culture-specific (subjective/relative) considerations, such as *morals* and *dogma*; in *NexGen Systemology* (*Grade-IV Metahuman Systemology*), a dynamic philosophy (applying "logic-and-reason") to understand the nature of "reality agreements" concerning *rightness* and *wrongness*, then treating the most optimum conditions of continued existence ("SURVIVAL" in *Beta-existence*; "CREATION" in *Alpha*) for the highest affected "Sphere of Existence" (on the *Standard Model*).

ethics processing : a *systematic processing* methodology introduced for bridging *Grade-IV Metahuman Systemology* (*Wizard Level-0*) with *Grade-V Spiritual Systemology* (*Wizard Level-1*) that emphasizes personal realization of "*Ethics*" and increased ability and responsibility to confront the "rightness" and "wrongness" of past actions (on the Backtrack), including defragmentation of "*Harmful Acts*" (as *Imprinting Incidents*) and any corresponding "*Hold-Backs*" and "*Hold-Outs*" (which reduce *Actualized Awareness* and prompt an individual to *withdraw* their *reach*); also "*Route-3E.*"

etymology : the origins of "words" and their development.

evaluate : to determine, assign or fix a set value, amount or meaning.

exacting : a demanding rigid effort to draw forth from.

executable : the supreme authoritative ability to carry out according to design.

existence : the *state* or fact of *apparent manifestation*; the resulting combination of the Principles of Manifestation: consciousness, motion and substance; continued *survival*; that which independently exists; the *'Prime Directive'* and sole purpose of all manifestation or Reality; the highest common intended motivation driving any *"Thing"* or *Life*.

existential : pertaining to existence, or some aspect or condition of existence.

exoteric : public knowledge or common understanding; the level of understanding and *Knowing* maintained by the "masses"; how a thing is generally understood "by all" or the opposite of *esoteric*.

experiential data : accumulated reference points we store as memory concerning our "experience" with Reality.

exponent : a person that is a critical example of something.

extant : in existence; existing.

exterior : outside of; on the outside; in *NexGen Systemology*, we mean specifically the POV of *Self* that is *'outside of'* the *Human Condition,* free of the physical and mental trappings of the Physical Universe; a metahuman range of consideration; see also *'Zu-Vision'*.

external : a force coming from outside; information received from outside sources; in *NexGen Systemology*, the objective *'Physical Universe'* existence, or *beta-existence*, that the Physical Body or *genetic vehicle* is essentially *anchored* to for its considerations of locational space-time as a dimension or POV.

extrapolate : to make an estimate of the "value" outside of the perceivable range.

extropy : *NexGen Systemology NewSpeak*—the reduction of organized spiritual systems back into a singularity of Infinity when their integrity is measured against space over time; reduction toward an infinitude; the opposite of *entropy*.

—**F**—

facets : an aspect, an apparent phase; one of many faces of something; a cut surface on a gem or crystal; in *NexGen Systemology*—a single perception or aspect of a memory or *"Imprint"*; any one of many ways in which a memory is recorded; perceptions associated with a painful emotional (sensation) experience and "*imprinted*" onto a metaphoric lens through which to view future similar experiences; other secondary terminals that are associated with a particular terminal, painful event or experience of loss, and which may exhibit the same encoded significance as the activating event.

faculties : abilities of the mind (individual) inherent or developed.

fallacy : a deceptive, misleading, erroneous and/or false beliefs; unsound logic; persuasions, invalidation or enforcement of Reality agreements based on authority, sympathy, bandwagon/mob mentality, vanity, ambiguity, suppression of information, and/or presentation of false dichotomies.

fate : what is brought to light or actualized as experience; the actual *course* taken to reach an end, charted end, or final *destination*; in *NexGen Systemology*, the *'fate'* of a '*Human Spirit*' (or '*Alpha Spirit'*) is determined by the choice of course taken to experience *Life*.

feedback loop : a complete and continuous circuit flow of energy or information directed as an output from a source to a target which is altered and return back to the source as an input; in *General Systemology*—the continuous process where outputs of a system are routed back as inputs to complete a circuit or loop, which may be closed or connected to other systems/circuits; in *NexGen Systemology*—the continuous process where directed *Life* energy and *Awareness* is sent back to *Self* as experience, understanding and memory to complete an energetic circuit as a loop.

flattening a wave : see "*process-out*" for definition; also see "*collapsing a wave.*"

flow : movement across (or through) a channel (or conduit); a direction of active energetic motion typically distinguished as either an *in-flow*, *out-flow* or *cross-flow*.

fodder : food, esp. for cattle; the raw material used to create.

forgive(ness) : to let go of resentment (against an offender, source of *Harmful-Act*) or give up emotional (energetic) turbulence connected to inclinations to punish; a legal pardon; to intentionally "overlook" (as opposed to "forget") the repayment of a debt or sense of something owed.

fractal : a wave-curve, geometric figure, form or pattern, with each part representative of the same characteristics as the whole; any baseline, sequence or pattern where the 'whole' is found in the 'parts' and the 'parts' contain the 'whole'; a pattern that reoccurs similarly at various scales/levels on a continuous whole; a subset of a Euclidean space explored in higher-level academic mathematics, in which fractal dimensions are found to exceed topological ones; in NexGen Systemology, a "fractal-like" description is used specifically for a pattern or form that has a reoccurring nature without regard to what level or scale it is manifest upon. Examples include the formation of crystals, tree-like patterns, the comparison of atoms to solar systems to galaxies, &tc.

fragmentation : breaking into parts and scattering the pieces; the *fractioning* of wholeness or the *fracture* of a holistic interconnected *alpha* state, favoring observational *Awareness* of perceived connectivity between parts; *discontinuity*; separation of a totality into parts; in *NexGen Systemology*, a person outside a state of *Self-Honesty* is said to be *fragmented*.

—**G**—

game : a strategic situation where a "player's" power of choice is employed or affected; a parameter or condition defined by purposes, freedoms and barriers (rules).

game theory : a mathematical theory of logic pertaining to strategies of maximizing gains and minimizing loses within prescribed boundaries and freedoms; a field of knowledge widely applied to human problem solving and decision-making; the application of true knowledge and logic to deduce the correct course of action given all variables and interplay of dynamic systems; logical study of decision making where "players" make choices that affect (the interests) of other "players"; an intellectual study of conflict and cooperation.

general systemology ("systematology") : a methodology of analysis and evaluation regarding the systems—their design and function; organizing systems of interrelated information-processing in order to perform a given function or pattern of functions.

genetic memory : the evolutionary, cellular and genetic (DNA) "memory" encoded into a *genetic vehicle* or *living organism* during its progression and duplication (reproduction) over millions (or billions) of years on Earth; in *NexGen Systemology*—the past-life Earth-memory carried in the genetic makeup of an organism (*genetic vehicle*) that is *independent of any* actual "spiritual memory" maintained by the *Alpha Spirit* themselves, from its own previous lifetimes on Earth and elsewhere using other *genetic vehicles* with no direct evolutionary connection to the current physical form in use.

genetic-vehicle : a physical *Life*-form; the physical (*beta*) body that is animated/controlled by the (*Alpha*) *Spirit* using a continuous *Lifeline* (ZU); a physical (*beta*) organic receptacle and catalyst for the (*Alpha*) *Self* to operate "causes" and experience "effects" within the *Physical Universe*.

gifted : attributing a special quality or ability; having exceptionally high intelligence or mental faculties.

gnosis : a *Greek* word meaning knowledge, but specifically "true knowledge"; the highest echelon of "true knowledge" accessible (or attained) only by mystical or spiritual faculties whereby actualized realizations are achieved independent of specialized education.

Gnostics : a name meaning "having knowledge" in Greek language (see also *gnosis*); an early sect of Judeo-Christian mysticism from the 1st Century AD emphasizing true knowledge by *Self-Honest* experience of metahuman and spiritual states of beingness, emphasizing defragmentation of "illusion" and overcoming of material "deception"; an esoteric proto-Systemology organization disbanded by the Roman Church as heretical.

godhood : a divine character or condition; "divinity."

gradient : a degree of partitioned ascent or descent along some scale, elevation or incline; "higher" and "lower" values in relation to one another.

GSR-Meters ("galvanic skin response"–"electropsychometer") : a *biofeedback* device used for measuring electrical resistance (in "Ohms") of the skin surface; one of many parts used in a polygraph system; a highly sensitive "Ohm-meter" with variable range, set points and amplification used to monitor electrical fluctuations of the skin surface.

—H—

harmful-act : a counter-survival mode of behavior or action (esp. that causes harm to one of more *Spheres of Existence*)—or—an overtly aggressive (hostile and/or destructive) action against an individual or any other *Sphere of Existence*; in *Utilitarian Systemology*—a shortsighted (serves fewest/lowest *Spheres of Existence*) intentional overtly harmful action to resolve a perceived problem; a revision of the rule for standard *Utilitarianism* for Systemology to distinguish actions which provide the least benefit to the least number of *Spheres of Existence*, or else the greatest harm to the greatest number of *Spheres of Existence*; in *moral philosophy*—an action which can be experienced by few and/or which one would not be willing to experience for themselves (*theft, slander, rape, &tc*); an iniquity or iniquitous act.

help : to assist survival of; aid continuing optimum success.

heralded : proclaimed ahead of or prior to; officially announced.

hold-back : withheld communications (esp. actions) such as "*Hold-Outs*"; intentional (or automatic) withdrawal (as opposed to reach); Self-restraint (which may eventually be enforced or automated); not reaching, acting or expressing, when one should be; an ability that is now restrained (on automatic) due to inability to withhold it on Self-determinism alone.

hold-outs : in photography, the numerous snapshots/pictures withheld from the final display or professional presentation of the event; withheld communications; in Utilitarian Systemology—energetic withdrawal and communication breaks with a "*terminal*" and its *Sphere of Existence* as a result of a "*Harmful-Act*"; unspoken or undiscovered (hidden, covert) actions that an individual withholds communications of, fearing punishment or endangerment of *Self-preservation* (*First Sphere*); the act of hiding (or keeping hidden) the truth of a "*Harmful-Act*"; a refusal to communicate with a *Pilot*; also "*Hold-Back.*"

holistic : the examination of interconnected systems as encompassing something greater than the *sum* of their "parts."

Homo Novus : literally, the "new man"; the "newly elevated man" or "known man" in ancient Rome; the man who "knows (only) through himself"; in NexGen Systemology—the next spiritual and intellectual evolution of *homo sapiens* (the "modern Human Condition"), which is signified by a demonstration of higher faculties of *Self-Actualization* and clear *Awareness*.

Homo Sapiens Sapiens : the present standard-issue Human Condition; the *hominid* species and genetic-line on Earth that received modification, programming and conditioning by the *Anunnaki* race of *Alpha-Spirits*, of which early alterations contributed to various upgrades (changes) to the genetic-line, beginning approximately 450,000 years ago (*ya*) when the *Anunnaki* first appear on Earth; a species for the Human Condition on Earth that resulted from many specific *Anunnaki* "genetic" and "cultural" *interventions* at certain points of significant advancement—specifically (but not limited to) *circa* 300,000 *ya,* 200,000 *ya,* 40,000 *ya,* and 8,000 *ya*; a species of the Human Condition set for replacement by *Homo Novus.*

hostile-motivation : an *imprint* of a counter-survival action (or "*Harmful-Act*") committed by another against Self, stored as data to justify future actions (retaliation, *&tc.*); any *Sphere of Existence* (though usually an individual) receiving the effect of a "*Harmful-Act*"; an *imprint* used to rationalize "motivation" or "justification" for committing a "*Harmful-Act*"; in systematic *games theory* —the *modus operandi* concerning "payback," "revenge" and "tit-for-tat."

hot button : something that triggers or incites an intense emotional reaction instantaneously; in *NexGen Systemology*—a slang term denoting a highly reactive *channel,* heavily *charged* with a long chain of cumulative *emotional imprinting,* typically (but not necessarily) connected to a significant or "primary" *implant*; a non-technical label, first applied during *Grade-IV Professional Piloting "Flight School"* research sessions of Spring-Summer 2020, to indicate specific circuits, channels or terminals that cause a *Seeker* to immediately react with intense emotional responses, whether in general, directed to the *Pilot,* or even at effectiveness of processing.

Human Condition : a standard default state of Human experience that is generally accepted to be the extent of its potential identity (*beingness*)—currently treated as *Homo Sapiens Sapiens,* but which is scheduled for replacement by *Homo Novus.*

humanistic psychology : a field of academic psychology approaching a holistic emphasis on *Self-Actualization* as an individual's most basic motivation; early key figures from the 20th century include: Carl Rogers, Abraham Maslow, L. Ron Hubbard, William Walker Atkinson, Deepak Chopra and Timothy Leary (to name a few).

hypothetical : operating under the assumption a certain aspect actual "is."

—I—

identification : the association of *identity* to a thing; a label or fixed data-set associated to what a thing is; association "equals" a thing, the "equals" being key; an equality of all things in a group, for example, an "apple" identified with all other "apples"; the reduction of "I-AM"-*Self* from a *Spiritual Beingness* to an "identity" of some form.

identity : the collection of energy and matter—including memory —across a "*Spiritual Timeline*" that we consider as "I" of *Self*, but the "I" is an individual and not an identification with anything other than *Self* as *Alpha-Spirit.*

identity-system : the application of the *ZU-line* as "I"—the continuous expression of *Self* as *Awareness* across a "*Spiritual Timeline*"; see "*identity.*"

illuminated : to supply with light so as to make visible or comprehensible.

imagination : the ability to create *mental imagery* in one's Personal Universe at will and change or alter it as desired; the ability to create, change and dissolve mental images on command or as an act of will; to create a mental image or have associated imagery displayed (or "conjured") in the mind that may or may not be treated as real (or memory recall) and may or may not accurately duplicate objective reality; to employ *Creative Abilities* of the Spirit that are independent of reality agreements with beta-existence.

Imaginomicon : the fourth professional publication of Mardukite Systemology, released publicly in mid- 2021; the second professional text in Grade-IV Metahuman Systemology, released as "*Liber-3D*"; contains fundamental theory of "*Spiritual Ability*" and "*Route-0*" systematic processing methodology.

immersion : plunged or sunk into; wholly surrounded by.

imperative : a high-level authoritarian command; a command triggering urgency and necessity of a certain goal or directive; see also "*Spheres of Existence*" and "*Prime Directive.*"

implant : to graft or surgically insert; to establish firmly by setting into; to instill or install a direct command or consideration in consciousness (Mind-System, &tc.); a mechanical device inserted beneath the surface/skin; in *Metahuman Systemology*, an "energetic mechanism" (linked to an Alpha-Spirit) composing a circuit-network and systematic array of energetic receptors underlying and filter-screening communication channels between the Mind-System and *Self*; an energetic construct installed upon entry of a Universe; similar to a platen or matrix or circuit-board, where each part records a specific type or quality of *emotionally encoded imprints* and other "heavily charged" *Mental Images* that are "impressed" by future encounters; a basic platform on which certain *imprints* and *Mental Images* are encoded (keyed-in) and stored (often beneath the surface of "knowing" or *Awareness* for that individual, although an implanted "command" toward certain inclinations or behavioral tendencies may be visibly observable.

imprint : to strongly impress, stamp, mark (or outline) onto a softer 'impressible' substance; to mark with pressure onto a surface; in *NexGen Systemology*, the term is used to indicate permanent Reality impressions marked by frequencies, energies or interactions experienced during periods of emotional distress, pain, unconsciousness, loss, enforcement, or something antagonistic to physical (personal) survival, all of which are are stored with other reactive response-mechanisms at lower-levels of *Awareness* as opposed to the active memory database and proactive processing center of the Mind; an experiential "memory-set" that may later resurface—be triggered or stimulated artificially—as Reality, of which similar responses will be engaged automatically; holographic-like imagery "stamped" onto consciousness as composed of energetic *facets* tied to the "snap-shot" of an experience.

imprinting incident : the first or original event instance communicated and *emotionally encoded* onto an individual's "*Spiritual Timeline*" (recorded memory from all lifetimes), which formed a permanent impression that is later used to mechanistically treat future contact on that channel; the first or original occurrence of some particular *facet* or mental image related to a certain type of *encoded response*, such as pain and discomfort, losses and victimization, and even the acts that we have taken against others along the Spiritual Timeline of our existence that caused them to also be *Imprinted*.

inadvertent : an unintended (knowingly) result caused by low-Awareness actions; applying effort (enacting change) outside Self-Honesty, leading to negligent oversights with harmful outcomes.

incarnation : a present, living or concrete form of some thing, idea or beingness; an individual lifetime or life-cycle from birth/creation to death/destruction independent of other lifetimes or cycles.

inception : the beginning, start, origin or outset.

incite : to urge on or cause; instigate; prove or stimulate into action.

indefinable : without a clear definition being currently presented.

individual : a person, lifeform, human entity or creature; a *Seeker* or potential *Seeker* is often referred to as an "individual" within

Mardukite Zuism and Systemology materials.

infinite existence : "immortality."

infinitude : being infinite; quantity or quality of *Infinity.*

inhibited : withheld, held-back, discouraged or repressed from some state.

iniquities : wickedness or wicked acts ("sinful" in religious use); literal etymology, "that which is not equal"; synonymous with *Harmful-Acts.*

"in phase" : see *"phase alignment."*

insistence : repeated use of a communicated energy into a form that demands acknowledgment, is more difficult to avoid or ignore.

institution : a social standard or organizational group responsible for promoting some system or aspect in society.

intention : the directed application of Will; to intend (have "in Mind") or signify (give "significance" to) for or toward a particular purpose; in *NexGen Systemology* (from the *Standard Model*)—the spiritual activity at WILL (5.0) directed by an *Alpha Spirit* (7.0); the application of WILL as "Cause" from a higher order of Alpha Thought and consideration (6.0), which then may continue to relay communications as an "effect" in the universe.

inter-dimensional : systems that are interconnected or correlated between the Physical Universe and the Spiritual Universe—or between "dimension states" observably identified as "physical," "emotional," "psychological" and "spiritual." The only point of true interconnectivity that we can systematically determine is called *"Life"* or the POV of *Self.*

interior : inside of; on the inside; in *NexGen Systemology*, we mean specifically the POV of *Self* that is fixed to the *'internal'* *Human Condition,* including the *Reactive Control Center* (RCC) and Mind-System or *Master Control Center* (MCC); within *beta-existence.*

intermediate : a distinct point between two points; actions between two points.

internal : a force coming from inside; information received from inside sources; in *NexGen Systemology*, the objective *'Physical Universe'* experience of *beta-existence* that is associated with the Physical Body or *genetic vehicle* and its POV regarding sensation and perception; from inside the body; within the body.

interrogation : obtaining specific information through responses to questions, such as in 'systematic processing' and other forms of two-way communication.

invalidate : decrease the level or degree or *agreement* as Reality.

invest : spend on; give or devote something in exchange for a beneficial result; to endow with.

—J—

justice : observable social actions (or consequential reaction) and predetermined civic (legal) processes employed in a society or group to uphold or enforce their reality agreements concerning "*law*"; a civic authority and administrative body responsible for carrying out practical/physical responses and penalties; the words, "*just*," "*justice*" and "*justification*," all stem from the Latin "*jus*" (meaning "*morally right*," "*law, in accordance with*" and "*lawful*") or "*iustus*" (expressing what is "true," "proper," "up-right" and "justified").

—K—

"kNow" : a creative spelling and use of semantics for "*know*" and "*now*" to indicate the state of present-time actualized "Awareness" as Self (Alpha-Spirit), developed for fun dual-meaning messages made by early Mardukite Systemologists in 2008-9, such as "Live in the kNow" or "Be in the kNow"—and even "Drown in the kNow" (parodying a song featuring Matisyahu, by electronic music duo, *Crystal Method*).

knowledge : clear personal processing of informed understanding; information (data) that is actualized as effectively workable understanding; a demonstrable understanding on which we may 'set' our *Awareness*—or literally a "know-ledge."

KI : an ancient cuneiform sign designating the *'physical zone'*; the *Physical Universe*—comprised of physical matter and physical energy in action across space and observed as time; a direction of motion toward material *Continuity*, away from or subordinate to the Spiritual (*'AN'*); the physical condition of existence providing for our *beta* state of *Awareness* experienced (and interacted with) as an individual *Lifeform* from our primary Alpha state of Identity or *I-AM-Self* in the *Spiritual Universe* ('*AN*').

kinetic : pertaining to energy of physical motion and movement.

—L—

law : a formal codified outline (or list) of *ethical* standards regarding social participation and acceptable behavior, like a "*code*," except that it *is* enforced by civic consequences (or even "*Cosmic Law*") when not adhered to, usually with punishment coming either by the group (exclusively) or by involvement with an "outside party" or societal (legal) authority; a predictable sequence of naturally occurring events that will consistently repeat under the right conditions (such as "*Cosmic Law*" or "*Natural Law*").

learned : highly educated; possessing significant knowledge.

level : a physical or conceptual *tier* (or plane) relative to a *scale* above and below it; a significant *gradient* observable as a *foundation* (or surface) built upon and subsequent to other levels of a totality or whole; a *set* of "*parameters*" with respect to other such *sets* along a *continuum*; in *NexGen Systemology*, a *Seeker's* understanding, *Awareness* as *Self* and the formal grades of material/instruction are all treated as "*levels.*"

Liber-One : First published in October 2019 as "*The Tablets of Destiny: Using Ancient Wisdom to Unlock Human Potential*" by Joshua Free; republished in the complete *Grade-III* anthology, "*The Systemology Handbook*"; revised in August 2022 as "*The Tablets of Destiny (Revelation): How Long-Lost Anunnaki Wisdom Can Change the Fate of Humanity.*"

Liber-Two : First published in October 2020 as "*Metahuman Destinations: Piloting the Course to Homo Novus*" by Joshua Free; an anthology of the *Grade-IV* "Professional Piloting Course," containing revised materials from *Liber-2C*, *Liber-2D* and (most of) *Liber-3C*; republished in the complete *Grade-IV* anthology, "*The Metahuman Systemology Handbook.*"

Liber-Three : see "*Liber-3E.*"

Liber-2B : First published in December 2019 as *"Crystal Clear: The Self-Actualization Manual & Guide to Total Awareness"* by Joshua Free; republished in the complete *Grade-III* anthology, "*The Systemology Handbook*"; revised in April 2022 as "*Crystal Clear (Handbook for Seekers): Achieve Self-Actualization and Spiritual Ascension in This Lifetime.*"

Liber-2C : First published in April 2020 as "*Communication and Control of Energy & Power: The Magic of Will & Intention (Volume One)*" by Joshua Free; revision republished as an integral part of the *Grade-IV* "Professional Piloting Course," in October 2020 within "*Metahuman Destinations*" (*Liber-Two*); republished in the complete *Grade-IV* anthology, "*The Metahuman Systemology Handbook.*"

Liber-2D : First published in June 2020 as "*Command of the Mind-Body Connection: The Magic of Will & Intention" (Volume Two)*" by Joshua Free; revision republished as an integral part of the *Grade-IV* "Professional Piloting Course," in October 2020 within "*Metahuman Destinations*" (*Liber-Two*); republished in the complete *Grade-IV* anthology, "*The Metahuman Systemology Handbook.*"

Liber-3C : First published in July 2020 as "*Now You Know: The Truth About Universes & How You Got Stuck in One*" by Joshua Free; a discourse in the *Grade-IV* Metahuman Systemology series; a revision of one part republished in October 2020 within the "*Professional Piloting Course*" manual, "*Metahuman Destinations*" (*Liber-Two*), a revision of the remaining part republished in June 2021 within the "*Imaginomicon*" (*Liber-3D*); republished in the complete *Grade-IV* anthology, "*The Metahuman Systemology Handbook.*"

Liber-3D : First published in June 2021 as "*Imaginomicon: The Gateway to Higher Universes (A Grimoire for the Human Spirit)*" by Joshua Free; a manual completing the *Grade-IV* (Metahuman Systemology) professional series with a treatment of "Wizard Level-0"; revised in June 2022 as "*Imaginomicon (Revised Edition): Approaching Gateways to Higher Universes (A New Grimoire for the Human Spirit)*"; republished in the complete *Grade-IV* anthology, "*The Metahuman Systemology Handbook.*"

Liber-3E (Liber-Three) : First published in April 2022 as "*The Way of the Wizard: Utilitarian Systemology (A New Metahuman Ethic)*" by Joshua Free; a professional manual bridging *Grade-IV* (Metahuman Systemology, *Wizard Level-0*) with *Grade-V* (Spiritual Systemology, *Wizard Level-1*); republished in the complete *Grade-IV* anthology, "*The Metahuman Systemology Handbook.*"

localized : brought together and confined to a particular place.

logic : philosophical science of correct *reasoning*.

logic equations : using symbols and basic mathematical logic to establish the validity of statements or to see how a variable within a system will change the result; a basic demonstration of proportion or relationship between variables in a system.

logistics : pertaining to the movement or transportation between locations.

—M—

macrocosmic : taking examples and system demonstrations at one level and applying them as a larger demonstration of a relatively higher level or unseen dimension.

malefactor : a person that knowingly commits *Harmful-Acts*; a source of frequent turbulence and destruction on a system.

manifestation : something brought into existence.

Marduk : founder of Babylonia; patron Anunnaki "god" of Babylon.

Mardukite Zuism : a Mesopotamian-themed (Babylonian-oriented) religious philosophy and tradition applying the spiritual technology based on *Arcane Tablets* in combination with "Tech" from *NexGen Systemology*; first developed in the New Age underground by Joshua Free in 2008 and realized publicly in 2009 with the formal establishment of the *Mardukite Chamberlains.* The text "*Tablets of Destiny*" is a cross-over from Mardukite Zuism (and Mesopotamian Neopaganism) toward higher spiritual applications of Systemology.

Master-Control-Center (MCC) : a perfect computing device to the extent of the information received from "lower levels" of sensory experience/perception; the proactive communication system of the "*Mind*"; a relay point of active *Awareness* along the Identity's *ZU-line*, which is responsible for maintaining basic *Self-Honest Clarity* of *Knowingness* as a *seat of consciousness* between the *Alpha-Spirit* and the secondary "*Reactive Control Center*" of a *Life-form* in *beta existence*; the Mind-center for an *Alpha-Spirit* to actualize cause in the *beta existence*; the analytical *Self-Determined* Mind-center of an *Alpha-Spirit used* to project *Will* toward the genetic body; the point of contact between *Spiritual Systems* and the *beta existence*; presumably the "*Third Eye*" of a being connected directly to the *I-AM-Self*, which is responsible for *determining* Reality at any time; in *NexGen Systemology*, this is plotted at (4.0) on the continuity model of the *ZU-line*.

"Master Grades" : literary materials by Joshua Free (written between 1995 and 2019) revised and compiled for the "Mardukite

Academy of Systemology" instructional grades—"Route of Magick & Mysticism" (*Grade I, Part A*), "Route of Druidism & Dragon Legacy" (*Grade I, Part D*), "Route of Mesopotamian Mysteries" (Grade II) and "Route of Mardukite Systemology" or "Pathway to Self-Honesty" (*Grade III*).

maxim : the greatest or highest *premise* of a paradigm or particular literary *treatment*; a concise rule for conducting action or treating some subject; the most relevant "proverbial adage" applicable.

MCC : see "*Master-Control-Center.*"

mental image : a subjectively experienced "picture" created and imagined into being by the Alpha-Spirit (or at lower levels, one of its automated mechanisms) that includes all perceptible *facets* of totally immersive scene, which may be forms originated by an individual, or a "facsimile-copy" ("snap-shot") of something seen or encountered; a duplication of wave-forms in one's Personal Universe as a "picture" that mirror an "external" Universe experience, such as an *Imprint*.

Mesopotamia : land between Tigris and Euphrates River; modern-day Iraq; the primary setting for ancient *Sumerian* and *Babylonian* traditions thousands of years ago, including activities and records of the *Anunnaki.*

metahumanism : an applied philosophy of *transhumanism* with an emphasis on "spiritual technologies" as opposed to "external" ones; a new state or evolution of the *Human Condition* achievable on planet Earth, rooted in *Self-Honesty*, whereby individuals are operating *exterior* to considerations that are fixed exclusively to the *genetic vehicle* (Human Body) and independent of the *emotional encoding* and *associative programming* typical of the present standard-issue *Human Condition.*

Metahuman Destinations : the third professional publication of Mardukite Systemology, released publicly in October 2020; the first professional text in Grade-IV Metahuman Systemology, released as "*Liber-Two*" and containing materials from *Liber-2C, Liber-2D* and *Liber-3C*; contains fundamental theory of "*Professional Piloting*" and "*Route-3*" systematic processing methodology. Reissued as two volumes in 2022.

meter : a device used to measure; see *GSR-Meter.*

methodology : a complete system of applications, methods, principles and rules to compose a '*systematic*' paradigm as a "whole"—

esp. a field of philosophy or science.

"mind's eye" : following semantics of archaic esoterica, the point where "mental pictures" (and senses) are generated that define what an individual believes they are experiencing in present time; activities or phenomenon described in archaic esoterica as the "Third-Eye" (or actualized MCC) where the *Alpha-Spirit* directly interacts with the organic *genetic vehicle* in *beta-existence*; in the semantics of basic Mardukite Zuism and Hermetic Philosophy, *Self-directed* activity on the plane of "mental consciousness" between "spiritual consciousness" of the *Alpha-Spirit* and "physical/emotional consciousness" of the *genetic vehicle*; *NexGen* 'slang' used to describe "consciousness activity" *Self-directed* by an actualized WILL.

misappropriated : put into use incorrectly; to apply ineffectively or as unintended by design or definition.

missed hold-out : an individual's *Hold-Out* that someone else nearly found out about, or which leaves the individual wondering if they did actually find out or not; undisclosed event when someone else's behavior or speech restimulates emotional-response-reactions ("worry" *&tc.*) about potential discovery of a withheld *Harmful-Act* or *Hold-Out*; in *systematic processing*, a Seeker's "held-out" (hidden) data that they expect to be discovered during a *session*, but which is *missed* by the Pilot.

morals : widely held culturally conditioned (socially learned) ethical standards of conduct used to "judge" *rightness* from *wrongness* of an individual's character, personality or actions (which may or may not be intellectually and emotionally influenced by "local" religious customs, taboos and *dogma*; basic social reality agreements determining "proper conduct" and "right actions" (behavior) based on civic *laws*, social *codes* and religious *doctrines* of a particular society or group and its own cultural experiences of *Reality.*

motor functions : internal mechanisms that allow a body to move.

—N—

Nabu : the *Anunnaki* "god of wisdom, writing and knowledge" for Babylonian (Mardukite) Tradition.

negligible : so small or trifle that it may be disregarded.

neophyte : a beginning initiate or novice to a particular sect or methodology; novitiate or entry-level grade of training, study and practice of an esoteric order or mystical lodge (fellowship).

neurotransmitter : a chemical substance released at a physiologic-al level (of the genetic vehicle) that bridges communication of ener-getic transmission between the *Mind-Body* systems, using the "nervous system" of the physical body; biochemical amino acids and peptides (neuropeptides), hormones, &tc.

NexGen Systemology : a modern tradition of applied religious philosophy and spiritual technology based on *Arcane Tablets* in combination with "*general systemology*" and "*games theory*" de-veloped in the New Age underground by Joshua Free in 2011 as an advanced futurist extension of the "*Mardukite Chamberlains*"; also referred to as "*Mardukite Systemology,*" "*Metahuman Systemology*" and "*Spiritual Systemology.*"

—O—

objective : concerning the "external world" and attempts to observe Reality independent of personal "subjective" factors.

occulted / to occult : hidden by or secreted away; to hide some-thing from view; otherwise *occlude*, to shut out, shut in, or block; to *eclipse*, or leave out of view.

one-to-one : see "*A-for-A.*"

optimum : the most favorable or ideal conditions for the best result; the greatest degree of result under specific conditions.

orchestration : to arrange or compose the performance of a system.

organic : as related to a physically living organism or carbon-based life form; energy-matter condensed into form as a focus or POV of Spiritual Life Energy (*ZU*) as it pertains to beta-existence of *this* Physical Universe (*KI*).

oscillation-alternation : a particular type of (or fluctuation) between two relative states, conditions or degrees; a wave-action between two degrees, such as is described in the action of the *pen-dulum effect*; a flux or wave-like energy in motion, across space, calculable as time; in systematic processing, alternation is the shift between two direction flows on a circuit channel, such as *inflow* and *outflow*, or between two types of processing, such as *objective* and *subjective*; alternation of a POV creates "space."

—P—

pantheism : religious philosophies that observe God as inherent within all aspects of the Physical Universe.

paradigm : an all-encompassing *standard* by which to view the world and *communicate* Reality; a standard model of reality-systems used by the Mind to filter, organize and interpret experience of Reality.

parameters : a defined range of possible variables with in a model, spectrum or continuum; the extent of communicable reach capable within a system or across a distance; the defined or imposed limitations placed on a system or the functions within a system; the extent to which a Life or "thing" can *be, do* or *know* along any channel within the confines of a specific system or spectrum of existence.

paramount : the most important; of utmost importance; "above all else."

participation : being part of the action; affecting the result.

patter : fast-talk; a manner of quickly delivered speech/words, esp. used to persuade or sell something.

patterns (probability patterns) : observation of cycles and tendencies to predict a causal relationship or determine the actual condition or flow of dynamic energy using a holistic systemology to understand Life, Reality and Existence as opposed to isolating or excluding perceived parts as being mutually separate from other perceived parts.

patron god : the most sacred deity of a region or city, of which most temples and religious services are directed; the personal deity of an individual.

PCL : see "*processing command line.*"

perception : internalized processing of data received by the *senses*; to become *Aware of* via the senses.

personality (program) : the total composite picture an individual "identifies" themselves with; the accumulated sum of material and mental mass by which an individual experiences as their timeline; a "beta-personality" is mainly attached to the identity of a particular physical body and the total sum of its own genetic memory in combination with the data stores and pictures maintained by the Alpha Spirit; a "true personality" is the Alpha Spirit as Self completely defragmented of all erroneous limitations and barriers to consideration, belief, manifestation and intention.

perturbation : the deviation from a natural state, fixed motion, or orbit system caused by another external system; disturbing or disquieting the serenity of an existent state; inciting observable apparent action using indirect or outside actions or 'forces'; the introduction of a new element or facet that disturbs equilibrium of a standard system; the "butterfly effect"; in *NexGen Systemology*, *'perturbation'* is a necessary condition for the *ZU-line* to function as a *Standard Model* of actual *'monistic continuity'*—which is a *Life-force* singularity expressed along a spectrum with potential interactions at each degree from any source; the influence of a degree in one state by activities of another state that seem independent, but which are actually connected directly at some higher degree, even if not apparently observed.

phase (identification) : in *NexGen Systemology*, a pattern of personality or identity that is assumed as the POV from *Self*; personal identification with artificial "personality packages"; an individual assuming or taking characteristics of another individual (often unknowingly as a response-mechanisms); also *"phase alignment."*

phase alignment or *"in phase"* : to be in synch or mutually synchronized, in step or aligned properly with something else in order to increase the total strength value; in *NexGen Systemology*, alignment or adjustment of *Awareness* with a particular identity, space or time; perfect *defragmentation* would mean being "in phase" as *Self* fully conscious and Aware as an Alpha-Spirit *in* present *space* and *time*, free of synthetic personalities.

philanthropy : charitable; the intention (or programmed desire) to generously provide personal wealth and service to the well-being and continued existence of others.

physics : regarding data obtained by a material science of observable motions, forces and bodies, including their apparent interaction, in the Physical Universe (specific to this *beta-existence*).

physiology : a material science of observable biological functions and mechanics of living organisms, including codification and study of identifiable parts and apparent systematic processes (specific to agreed upon makeup of the *genetic vehicle* for this *beta-existence*).

pilfering : to steal in small quantities; petty theft.

pilot : a professional steersman responsible for healthy functional operation of a ship toward a specific destination; in *NexGen Systemology*, an intensive trained individual qualified to specially apply *Systemology Processing* to assist other *Seekers* on the *Pathway*.

ping : a short, high pitched ring, chime or noise that alerts to the presence of something; in computer systems, a query sent on a network or line to another terminal in order to determine if there is a connection to it; in *NexGen Systemology*, the sudden somatic twinge or pain or discomfort that is felt as a sensation in the body when a particular terminal (lifeform, object, concept) is 'brought to mind' or contacted on a personal communication channel-circuit; the accompanying sensations and mental images that are experienced as an automatic-response to the presence of some channel or terminal.

player (game theory) : an individual that is making decisions in a game and/or is affected by decisions others are making in the game, especially if those other-determined decisions now affect the possible choices.

point-of-view (POV) : a point to view from; an opinion or attitude as expressed from a specific identity-phase; a specific standpoint or vantage-point; a definitive manner of consideration specific to an individual phase or identity; a place or position affording a specific view or vantage; circumstances and programming of an individual that is conducive to a particular response, consideration or belief-set (paradigm); a position (consideration) or place (location) that provides a specific view or perspective (subjective) on experience (of the objective).

postulate : to put forward as truth; to suggest or assume an existence *to be*; to state or affirm the existence of particular conditions; to provide a basis of reasoning and belief; a basic theory accepted as fact; in *NexGen Systemology*, "Alpha-Thought"—the top-most decisions or considerations made by the Alpha-Spirit regarding the *"is-ness"* (what things "are") about energy-matter and space-time.

potentiality : the total "sum" (collective amount) of "latent" (dormant—present but not apparent) capable or possible realizations; used to describe a state or condition of what has not yet manifested, but which can be influenced and predicted based on observed patterns and, if referring to beta-existence, Cosmic Law.

POV : see *"point-of-view"* and/or *"POV Processing."*

POV processing : a methodology of *Grade-IV Metahuman Systemology* emphasizing systematic processing toward realizations that improve a Seeker's willingness to manage a present POV and associated *phases*, their ability to transfer POVs freely, increased tolerance to experiences (or encounters) with any other viewpoint, and finally, an actualized realization that a POV is not one-to-one with *Beingness* of *Self*; an extension of *creativeness processing* and "Wizard Level" training that systematically handles *Awareness* of "points" and "spots" in space, from which an Alpha-Spirit may place its own viewpoint of a dimension or Universe—also a pre-requisite to upper-route practices such as "*Zu-Vision*" and "*Back-track.*"

precedent : a matter which precedes or goes before another in importance.

precipitate : to actively hasten or quicken into existence.

preconception : to assign values or evaluate a reaction or response to a past "imprint" of something and treat it as present knowledge or experience.

prehistoric : any time before human history is properly recorded in writing; prior to c. 4000 B.C.

premise : a basis or statement of fact from which conclusions are drawn.

presence : the quality of some thing (energy/matter) being "present" in space-time; personal orientation of *Self* as an *Awareness* (*POV*) located in present space-time (environment) and communicating with extant energy-matter.

prevalent : of wide extent; an extensive or largely accepted aspect or current state.

Prime Directive : a "spiritual" implant program that installs purposes and goals into the personal experience of a Universe, esp. any *Beta-Existence* (whether a 'Games Universe' or a 'Prison Universe'); intellectually treated as the "Universal Imperative" in some schools of moral philosophy; comparable to "Universal Law" or "Cosmic Ordering."

probability : the causal likelihood for something to result, "effect" or manifest in and as a certain way, manner or degree, based on "observed evaluation" of programming and tendencies that follow Cosmic Law.

"process-out" or **"flatten a wave"** : to reduce *emotional encoding* of an *imprint* to zero; to dissolve a *wave-form* or *thought-formed* "solid" such as a "*belief*"; to completely run a *process* to its end, thereby *flattening* any previously "*collapsed-waves*" or *fragmentation* that is obstructing the *clear channel* of *Self-Awareness*; also referred to as "processing-out"; to discharge all previously held emotionally encoded imprinting or erroneous programming and beliefs that otherwise fix the free flow (wave) to a particular pattern, solid or concrete "*is*" form.

processing, systematic : the inner-workings or "through-put" result of systems; in *NexGen Systemology*, a methodology of applied spiritual technology used toward personal Self-Actualization; methods of selective directed attention, communicated language and associative imagery that targets an increase in personal control of the human condition.

processing command line (PCL) or **command line** : a directed input; a specific command using highly selective language for *Systemology Processing*; a predetermined directive statement (cause) intended to focus concentrated attention (effect).

projecting awareness : sending out (motion) or radiating "*consciousness*" from *Self* ("I") to another POV.

proportional : having a direct relationship or mutual interaction with.

protest : a response-communication objecting an enforcement or a rejection of a prior communication; an effort to cancel, rewrite or destroy the existence or "is-ness" (what something "is") of a previous creation or communication; unwillingness to be the Point-of-View of effect or (receipt-point) for a communication.

Proto-Indo-European (PIE) : in Linguistic-Semantic Sciences, a hypothetical single-source Eurasian root language (c.4500 B.C.) demonstrating common origins of many "word-roots" found in European languages.

psychokinesis (PK) / telekinesis : influencing a (physical) system without (physical) interaction; *psychokinesis* from the Greek for 'soul' and 'movement', and *telekinesis* from the Greek for 'at a distance' and 'movement'.

psychometric evaluation : the relative measurement of personal ability, mental (psychological/thought) faculties, and effective processing of information and external stimulus data; a scale used in "applied psychology" to evaluate and predict human behavior.

—R—

rationality / reasoning (game theory) : the extent to which a player seeks to play (make decisions, &tc.) in order to maximize the gains (or else survival) achievable within any given game conditions; the ability and willingness of an individual to reach toward conditions that promote the highest level of survival and existence and make the best choices and moves to see the desired goal manifest.

reactive control center (RCC) : the secondary (reactive) communication system of the "*Mind*"; a relay point of *Awareness* along the Identity's *ZU-line*, which is responsible for engaging basic motors, biochemical processes and any *programmed automated responses* of a living *beta* organism; the reactive Mind-Center of a living organism relaying communications of *Awareness* between causal experience of *Physical Systems* and the "*Master Control Center*"; it presumably stores all emotional encoded imprints as fragmentation of "chakra" frequencies of *ZU* (within the range of the "*psychological/emotive systems*" of a being), which it may *react* to as Reality at any time; in *NexGen Systemology*, this is plotted at (2.0) on the continuity model of the *ZU-line*.

reality : see "*agreement.*"

realization : the clear perception of an understanding; a consideration or understanding on what is "actual"; to make "real" or give "reality" to so as to grant a property of "beingness" or "being as it is"; the state or instance of coming to an *Awareness*; in *NexGen Systemology*, "gnosis" or true knowledge achieved during *systematic processing*; achievement of a new (or "higher") cognition, true knowledge or perception of Self; a consideration of reality or assignment of meaning.

receptacle : a device or mechanism designed to contain and store a specific type of aspect or thing; a container meant to receive something.

recursive : repeating by looping back onto itself to form continuity; *ex.* the "Infinity" symbol is recursive.

relative : an apparent point, state or condition treated as distinct from others.

religion : a concise spiritual *paradigm*, set of beliefs and practices regarding "Divinity," "Infinite Beingness"—or else, "God"—as representative symbol of the *Eighth Sphere of Existence* for *Beta-Existence* (or else "Infinity").

relinquish : to give up control, command or possession of.

repetitively : to repeat "over and over" again; or else "repetition."

responsibility : the *ability* to *respond*; the extent of mobilizing *power* and *understanding* an individual maintains as *Awareness* to enact *change*; the proactive ability to *Self-direct* and make decisions independent of an outside authority.

resurface : to return to (or bring up to) the "surface" of that which has previously been submerged; in *NexGen Systemology*—relating specifically to processes where a *Seeker* recalls blocked energy stored covertly as emotional "*imprints*" (by the RCC) so that it may be effectively defragmented from the "*ZU-line*" (by the MCC).

rhetoric : the art, study or skilled craft of using language eloquently (words, writing, speech preparation); expert communication using "words"; effectively using language for persuasive communication.

Route-0 : a specific methodology from *SOP-2C* denoting "*Creativeness Processing*," as described in the text "*Imaginomicon*" (*Liber-3D*).

Route-0E : a specific methodology (expanding on *Route-0* from *Liber-3D*) denoting "*Conceptual Processing*" applied to *Ethics Beta-Defragmentation*, as described in the text "*Way of the Wizard*" (*Liber-Three* or *Liber-3E*).

Route-1 : a specific methodology from *SOP-2C* denoting "*Resurfacing Processing*," as described in the text "*Tablets of Destiny*" (*Liber-One*) as "RR-SP" (and reissued in "*The Systemology Handbook*").

Route-2 : a specific methodology from *SOP-2C* denoting "*Analytical-Recall Processing*," as described in the text "*Crystal Clear*" (*Liber-2B*) as "AR-SP" (and reissued in "*The Systemology Handbook*").

Route-3 : a specific methodology from *SOP-2C* denoting "*Communication-Circuit Processing*," as described in the text "*Metahuman Destinations*" (*Liber-Two*); also the basis for *SOP-2C* routine.

Route-3E : a specific methodology (expanding on *Route-3* from *SOP-2C*) denoting "*Ethics Processing*," as described in the text "*The Way of the Wizard*" (*Liber-Three* or *Liber-3E*); also related to "Standard Procedure R-3E."

—S—

science : a systematized *paradigm* of Knowingness—from the Latin '*scire*', meaning "know"; an empirical and objective understand-

ing of data collected by observation, calculation and logical deduction—and which may usually be used to predict phenomenon or occurrences in the Physical Universe ("*Beta-Existence*").

scions : a descendant or offspring; an offshoot or branch.

Seeker : an individual on the *Pathway to Self-Honesty*; a practitioner of *Mardukite Systemology* or *NexGen Systemology Processing*
that is working toward *Spiritual Ascension.*

Self-actualization : bringing the full potential of the Human spirit
into Reality; expressing full capabilities and creativeness of the *Alpha-Spirit.*

Self-determinism : the freedom to act, clear of external control or
influence; the personal control of Will to direct intention.

Self-evaluation : see "*psychometric evaluation.*"

Self-honesty : the basic or original *alpha* state of *being* and *knowing*; clear and present total *Awareness* of-and-as *Self*, in its most basic and true proactive expression of itself as *Spirit* or *I-AM*—free of
artificial attachments, perceptive filters and other emotionally-react-
ive or mentally-conditioned programming imposed on the human
condition by the systematized physical world; the ability to experience existence without judgment.

self-sustained : self-supported; self-sufficient; independent.

semantics : the *meaning* carried in *language* as the *truth* of a
"thing" represented, *A-for-A*; the *effect* of language on *thought*
activity in the Mind and physical behavior; language as *symbols*
used to represent a concept, "thing" or "solid."

semantic-set : the implied meaning behind any groupings of words
or symbols used to define a specific paradigm.

sensation : an external stimulus received by internal sense organs
(receptors/sensors); sense impressions.

sentient : a living organism with consciousness or intelligence; a
"thinking" or "reasoning" being that perceives information from the
"senses."

simulacrum : an tangible likeness, image, facsimile or superficial
representation that is similar to or resembles someone or something
else; in *NexGen Systemology*, any *genetic vehicle* or physical body
is considered a reflective "simulacrum" of, and used as a "ves-
sel-shell" by, the *Alpha-Spirit* or *Self* (I-AM), which otherwise

maintains no true finite locatable form in *beta-existence*.

sine-wave : the *frequency* and amplitude of a quantified (calculable) *vibration* represented on a graph (graphically) as smooth repetitive *oscillation* of a *waveform*; a *waveform* graphed for demonstration—otherwise represented in *NexGen Systemology* logic equations as 'Wf,' or in mathematics as the *'function of x'* (fx); graphically representing arcs (*parameters*) of a circular *continuity* on a *continuum*; in the *Standard Model of NexGen Systemology*, the actual 'wave vibration' graphically displayed on an otherwise static *ZU-line* (of Infinity) is a *'sine-wave'*.

singularity : in general use, "to be singular," but our working definition suggests the opposite of individuality (contrary to most dictionaries); in upper-level sciences, a "zero-point" where a particular property or attribute is mathematically treated as "infinite" (such as the "black-hole" phenomenon), or else where apparently dissimilar qualities of all existing aspects (or individuals) share a "singular" expression, nature or quality; additionally, in *NexGen Systemology*, a hypothetical zero-point when apparent values of all parts in a Universe are equal to all other parts before it collapses; in *Transhumanism*, a hypothetical "runaway reaction" in technology, when it becomes self-aware, self-propagating, self-upgradable and self-sustainable, and replaces human effort of advancement or even makes continued human existence impossible; also, technological efforts to maintain an artificial immortality of the Human Condition on a digital mainframe.

slate : a hard thin flat surface material used for writing on; a chalkboard, which is a large version of the original wood-framed writing slate, named for the rock-type it was made from.

somatic : specifically pertaining to the physical body, its sensations and response actions or behaviors as separate from a "Mind-System"; also *"pings."*

SOP-2C : *Standard Operating Procedure #2C* or *Systemology Operating Procedure #2C*; a standardized procedural formula introduced in materials for *"Metahuman Destinations"* (*Liber-Two*); a regimen or outline for standard delivery of systematic processing used by *Systemology Pilots* and *Mardukite Ministers*; a procedure outline of systematic processing, which includes applications of *"Route-1," "Route-2," "Route-3"* and *"Route-0"* as taught for *Grade-IV Professional Piloting*.

space : a viewpoint or *Point-of-View* (POV) extended from any point out toward a dimension or dimensions; the consideration of a point or spot as an *anchor* or *corner* in addition to others, which collectively define parameters of a dimensional plane; the field of energy/matter mass created as a result of communication and control in action and measured as time (wave-length), such as "distance" between points (or peaks on a wave).

spectrum : a broad range or array as a continuous series or sequence; defined parts along a singular continuum; in physics, a gradient arrangement of visible colored bands diffracted in order of their respective wavelengths, such as when passing *White Light* through a *prism*.

Spheres of Existence (dynamic systems) : a series of *eight* concentric circles, rings or spheres (each larger than the former) that is overlaid onto the Standard Model of Beta-Existence to demonstrate the dynamic systems of existence extending out from the POV of Self (often as a "body") at the *First Sphere*; these are given in the basic eightfold systems as: *Self, Home/Family, Groups, Humanity, Life on Earth, Physical Universe, Spiritual Universe* and *Infinity-Divinity.*

spiritual timeline : a continuous stream of moment-to-moment *Mental Images* (or a record of experiences) that defines the "past" of a spiritual being (or *Alpha-Spirit*) and which includes impressions (*imprints, &tc.*) form all life-incarnations and significant spiritual events the being has encountered; in NexGen Systemology, also "*backtrack.*"

standard issue : equally dispensed to all without consideration.

standard model : a fundamental *structure* or symbolic construct used to evaluate a complete *set* in *continuity* relative to itself and variable to all other *dynamic systems* as graphed or calculated by *logic*.

Standard Model, The (systemology) : in *NexGen Systemology*— our existential and cosmological *standard model* or cabbalistic model; a "*monistic continuity model*" demonstrating *total system* interconnectivity "above" and "below" observation of any apparent *parameters*; the original presentation of the *ZU-line*, represented as a singular vertical (*y*-axis) waveform in space across dimensional levels or Universes (*Spheres of Existence*) without charting any specific movement across a dimensional time-graph *x*-axis; The Standard Model of Systemology represents the basic workable synthesis of common denominators in models explored throughout Grade-I and Grade-II material.

static : characterized by a fixed or stationary condition; having no apparent change, movement or fluctuation.

stoicism : pertaining to the school of "stoic" philosophy, distinguished by calm mental attitudes, freedom from desire/passion and essentially any emotional fluctuation.

sub-zones : at ranges "below" which we are representing or which is readily observable for current purposes.

successively : what comes after; forward into the future.

succumb : to give way, or give in to, a relatively stronger superior force.

Sumerian : ancient civilization of *Sumer,* founded in Mesopotamia c. 5000 B.C.

superfluous : excessive; unnecessary; needless.

superstition : knowledge accepted without good reason.

surefooted : proceeding surely; not likely to stumble or fall.

symbiotic : pertaining to the closeness, proximity and affinity between two beings that are in mutual communication or maintaining mutually validating interactions.

symbol : a concentrated mass with associated meaning or significance.

sympathy : a sensation, feeling or emotion—of anger, fear, sorrow and/or pity—that is a *personal reaction* to the misfortune and failure of another being.

syntax : from the Greek, "to arrange together"; the semantic meaning that words convey when combined together; the manner in which words are arranged together to provide an understandable meaning, such as following the structure for a sentence.

system : from the Greek, "to set together"; to set or arrange things or data together so as to form an orderly understanding of a "whole"; also a *'method'* or *'methodology'* as an orderly standard of use or application of such data arranged together.

systematization : to arrange into systems; to systematize or make systematic.

Systemology : see *"NexGen Systemology."*

Systemology Procedure 1-8-0 : advanced spiritual technology within our Systemology, which applies a methodology of systematic practice for experiencing: (1) Self-Awareness, (8) Nothingness and (0) Beingness, introduced for "Crystal Clear" but expanded on for "*Imaginomicon*"; 'one-eight-zero' is included in, but not the same as application 'one-eighty'—or else the *Beta-Defrag-Intensive* called "*SOP-180*" or "*Systemology-180.*"

Systemology-180 : an intensive systematic processing routine employing all *Grade-III, Grade-IV* and cross-over *Wizard-Level* work to date; the total sum of all effective philosophical and spiritual applications necessary to professionally *Pilot* a *Seeker* to reach a stable point of *Self-Honesty* and basic *Beta-Defragmentation*, as a prerequisite to treating "*Actualized-Ascension Technologies*" (*A.T.*) of upper-level *Wizard Grades*.

systems theory : see *"general systematology"*

—T—

Tablets of Destiny : the first professional publication of Mardukite Systemology, released publicly in October 2019; the first professional text in Grade-III Mardukite Systemology, released as "*Liber-One*" and reissued in the Grade-III Master Edition "*Systemology Handbook*"; contains fundamental theory of the "*Standard Model*" and "*Route-1*" systematic processing methodology; revised in 2022 as "*The Tablets of Destiny (Revelation).*"

telekinesis : see *"psychokinesis."*

teleological (teleology) : using the end-goal or purpose of something as an explanation of its function (rather than being a function of its cause); example—Aristotle wrote (in his discourse, "*Metaphysics*") that the intrinsic (inherent or true nature) *telos* of an 'acorn' is to become a fully formed 'oak tree'; the ends are an underlying purpose, not the cause (also known as "final cause"), or else the famous phrase: "the ends justify the means."

terminal (node) : a point, end or mass on a line; a point or connection for closing an electric circuit, such as a post on a battery terminating at each end of its own systematic function; any end point or 'termination' on a line; a point of connectivity with other points; in systems, any point which may be treated as a contact point of interaction; anything that may be distinguished as an 'is' and is therefore a 'termination point' of a system or along a flow-line which may interact with other related systems it shares a line with; a point of interaction with other points.

thought-experiment : from the German, *Gedankenexperiment*; logical *considerations* or mental models used to concisely visualize consequences (cause-effect sequences) within the context of an imaginary or hypothetical scenario; using faculties of the Mind's Eye to *Imagine* things accurately with *considerations* that *have not* already been consciously experienced in *beta-existence*.

thought-form : apparent *manifestation* or existential *realization* of *Thought-waves* as "solids" even when only apparent in Reality-agreements of the Observer; the treatment of *Thought-waves* as permanent *imprints* obscuring *Self-Honest Clarity* of *Awareness* when reinforced by emotional experience as actualized "thought-formed solids" ("*beliefs*") in the Mind; energetic patterns that "surround" the individual.

thought-habit : reoccurring modes of thought or repeated "self-talk"; essentially "self-hypnosis" resulting in a certain state.

thought-wave or **wave-form** : a proactive *Self-directed action* or reactive-response *action* of *consciousness*; the *process* of *thinking* as demonstrated in *wave-form*; the *activity* of *Awareness* within the range of *thought vibrations/frequencies* on the existential *Life-continuum* or *ZU-line*.

threshold : a doorway, gate or entrance point; the degree to which something is to produce an effect within a certain state or condition; the point in which a condition changes from one to the next.

thwarted : to successfully oppose or prevent a purpose from actualizing.

tier : a series of rows or levels, one stacked immediately before or atop another.

time : observation of cycles in action; motion of a particle, energy or wave across space; intervals of action related to other intervals of action as observed in Awareness; a measurable wave-length or frequency in comparison to a static state; the consideration of variations in space.

timeline : plotting out history in a linear (line) model to indicate instances (experiences) or demonstrate changes in state (space) as measured over time; a singular conception of continuation of observed time as marked by event-intervals and changes in energy and matter across space.

tipping point : a definitive "point" when a series of small changes (to a system) are significant enough to be *realized* or *cause* a larger, more significant change; the critical "point" (in a system) beyond which a significant change takes place or is observed; the "point" at which changes that cross a specific "threshold" reach a noticeably new state or development.

transhumanism : a social science and applied philosophy concerning the next evolved state of the "*Human Condition,*"; progress in two potential directions, either "spiritual" technologies advancing *Self* as an "Alpha-Spirit," or the direction of "external"-"physical" technologies that modify or eliminate characteristics of the *Body*; a theme describing contemporary application of material sciences emphasizing only "physical" and "genetic" parts of the *Human* experience, such as brain activity, cell-life extension and space travel; *NexGen Systemology* recently began distinguishing its emphasis on "spiritual technology" as "*metahumanism.*"

transmit : to send forth data along some line of communication; to move a point across a distance.

traumatic encoding : information received when the sensory faculties of an organism are "shocked" into learning it as an "emotionally" encoded *Imprint*; a duplicated facsimile-copy or *Mental Image* of severe misfortune, violent threats, pain and coercion, which is then categorized, stored and reactively retrieved based exclusively on its emotional *facets.*

treat / treatment : an act, manner or method of handling or dealing with someone, something or some type of situation; to apply a specific process, procedure or mode of action toward some person, thing or subject; use of a specific substance, regimen or procedure to make an existing condition less severe; also, a written presentation that handles a subject in a specific manner.

turbulence : a quality or state of distortion or disturbance that creates irregularity of a flow or pattern; the quality or state of aberration on a line (such as ragged edges) or the emotional "turbulent feelings" attached to a particular flow or terminal node; a violent, haphazard or disharmonious commotion (such as in the ebb of gusts and lulls of wind action).

—U—

unconscious : a state when *Awareness* as *Self* is removed totally from the equation of *Life* experience, though it continues to be recorded in lower-level response mechanisms (fixed to a simulacrum or genetic vehicle) for later retrieval.

undefiled : to remain intact, untouched or unchanged; to be left in an original "virgin" state.

understanding : a clear 'A-for-A' duplication of a communication as 'knowledge', which may be comprehended and retained with its significance assigned in relation to other 'knowledge' treated as a 'significant understanding'; the "grade" or "level" that a knowledge base is collected and the manner in which the data is organized and evaluated.

Utopian Philosophy : a social philosophy and ethic for (primarily) independent rural (country-dwelling or pagan) living communities that adopt a neo-Utilitarian moral philosophy (as suggested by Systemology) to enhance the "greater happiness" and "Ascension" of all participants.

—V—

validation : reinforcement of agreements or considerations as "real."

vantage : a point, place or position that offers an ideal viewpoint (POV).

Venn diagram : a diagram for symbolic logic using circles to represent sets and their systematic relationship; popularized by logician *John Venn.*

verbatim : precisely reproduced or duplicated communication *one-to-one* or "word"-for-"word" (*'A-for-A'*).

via : literally, "by way of"; from the Latin, meaning "way."

vibration : effects of motion or wave-frequency as applied to any system.

viewpoint : see *"point-of-view" (POV).*

vizier : a high ranking official; a minister-of-state.

—W—

wave-form : see *"sine-wave."*

wave-function collapse : see *"collapsing a wave."*

Western Civilization : modern contemporary culture, ideals, values and technology, particularly of Europe and North America as distinguished by growing urbanization, industrialization, and inspired by a history of rebellion to strong religious and political indoctrination.

will *or* **WILL** (5.0) : in *NexGen Systemology* (from the *Standard Model*), the Alpha-ability at "5.0" of a Spiritual Being (*Alpha Spirit*) at "7.0" to apply *intention* as "Cause" from consideration or Alpha-Thought at "6.0" that is superior to "beta-thoughts" that only manifest as reactive "effects" below "4.0" and *interior* to the *Human Condition*.

willingness : the state of conscious Self-determined ability and interest (directed attention) to *Be*, *Do* or *Have*; a Self-determined consideration to reach, face up to (*confront*) or manage some "mass" or energy; the extent to which an individual considers themselves able to participate, act or communicate along some line, to put attention or intention on the line, or to produce (create) an effect.

—**Z**—

ziggurat : religious temples of ancient Mesopotamia; stepped-pyramids used for spiritual and religious purposes by Sumerians and Babylonians, many of which are presented as seven tiers, levels or terraces representing "Seven Gates" (or "7 Veils") of existence, separating material continuity of the Earth Plane from "Infinity."

ZU : the ancient Sumerian cuneiform sign for the archaic verb—"*to know*," "*knowingness*" or "*awareness*"; in *Mardukite Zuism and Systemology*, the active energy/matter of the "Spiritual Universe" (AN) experienced as a *Lifeforce* or *consciousness* that imbues living forms extant in the "Physical Universe" (KI); "*Spiritual Life Energy*"; energy demonstrated by the WILL of an actualized *Alpha-Spirit* in the "Spiritual Universe" (AN), which impinges its *Awareness* into the Physical Universe (KI), animating/controlling *Life* for its experience of *beta-existence* along an individual Alpha-Spirit's personal *Identity-continuum*, called a *ZU-line*.

Zu-**Line** : a theoretical construct in *Mardukite Zuism and Systemology* demonstrating *Spiritual Life Energy* (ZU) as a personal individual "continuum" of Awareness interacting with all Spheres of Existence on the Standard Model of Systemology; a spectrum of potential variations and interactions of a monistic continuum or singular *Spiritual Life Energy (ZU)* demonstrated on the Standard

Model; an energetic channel of potential POV and "locations" of Beingness, demonstrated in early Systemology materials as an individual Alpha-Spirit's personal *Identity-continuum*, potentially connecting *Awareness (ZU)* of *Self* with *"Infinity"* simultaneous with all points considered in existence; a symbolic demonstration of the *"Life-line"* on which *Awareness (ZU)* extends from the direction of the "Spiritual Universe" (AN) in its true original *alpha state* through an entire possible range of activity resulting in its *beta state* and control of a *genetic-entity* occupying the *Physical Universe (KI)*.

Zu-Vision : the true and basic (*Alpha*) Point-of-View (perspective, POV) maintained by *Self* as *Alpha-Spirit* outside boundaries or considerations of the *Human Condition* "Mind-Systems" and *exterior* to beta-existence reality agreements with the Physical Universe; a POV of Self *as* "a unit of Spiritual Awareness" that exists independent of a "body" and entrapment in a *Human Condition*; "spirit vision" in its truest sense.

WOULD

YOU

LIKE

TO

KNOW

MORE

? ? ?

SYSTEMOLOGY

KEYS TO THE KINGDOM

Spiritual Clearing (Levels 7 and 8)

New Standard
Systemology
Advanced Training
(A.T.) Course

*developed by
Joshua Free*

All *eleven* lesson-manuals of the new standard
Advanced Training Course on Mardukite Systemology
are combined together in *two volumes* as
"Keys to the Kingdom."

Also available individually.

"The Secret of Universes"

"Games, Goals and Purposes"

"The Jewel of Knowledge"

"Implanted Universes"

"Systemology Biofeedback"

"Systemology Procedures"

...and many more!

New Workbook Editions Coming Soon in 2025!

IN A WORLD FULL OF "TENS" BE AN
ELEVEN

THE METAPHYSICS OF
STRANGER THINGS

TELEKINESIS, TELEPATHY & SYSTEMOLOGY

by Joshua Free

Mardukite Systemology Liber-011

Experimental exploratory edition

Discover the metaphysical truth about the Universe—and maybe even yourself—as we explore what lies beneath the epic saga, *Stranger Things.* You're invited to a world where fantasy, science fiction and horror unite, and games like *Dungeons and Dragons* become reality.

Uncover a world of secret "mind control" projects, just like those at *Hawkins National Laboratory*. Decades of psychedelic experiments among other developmental programs for psychic powers, remote viewing, telekinesis (psychokinesis, PK-power) and more are revealed. Get an inside look at the operations of a real-life underground organization pursuing the truth about rehabilitating spiritual abilities for an actual "metahuman" evolution on planet Earth.

Premiere Edition available in both paperback and hardcover!

Seekers and students of the *Basic Course* and *Professional Course* will also be interested in the *Systemology Core Research Volumes*. These eight volumes are a complete chronological record of the *Mardukite New Thought* developments from the *Systemology Society,* as presented to the public from 2019 through 2023.

Our *Systemology Core* begins with the first professional publication released when the *Mardukite Systemology Society* emerged into public view from the underground in 2019, with: *"The Tablets of Destiny Revelation."*

The Tablets of Destiny Revelation:
*How Long-Lost Anunnaki Wisdom
Can Change the Fate of Humanity*

Crystal Clear: *Handbook for Seekers*

Metahuman Destinations (*2 volumes*)

Imaginomicon:
Approaching Gateways to Higher Universes

Way of the Wizard: *Utilitarian Systemology*

Systemology-180: *Fast-Track to Ascension*

Systemology Backtrack:
Reclaiming Spiritual Power & Past-Life Memory

New Deluxe Oversized Hardcover Edition for 2023!

NOVEM PORTIS (DELUXE EDITION)
NECRONOMICON REVELATIONS,
NINE GATES OF THE KINGDOM OF SHADOWS
& CROSSING TO THE ABYSS
10th Anniversary—Deluxe Hardcover (*Liber-R, 9+555*)
Collected Works by Joshua Free

Commemorating completion of the "Necronomicon Shadows"
cycle of research and development by the Mardukite
Chamberlains (2009–2012). Originally intended as a
research-companion to "*Necronomicon: The Anunnaki Bible*"
and the remaining 'Core', a Mardukite anthology of this cycle of
work—known as "*Nine Gates*" or "*Novem Portis*"—
eventually developed into an underground bestseller by itself.

In addition to other bonus articles and supplements,
a complete collection of material from *Liber-9*, *Liber-R* and *Liber-555*
are together in a deluxe hardcover anthology edition
for the first time ever!

∞

A mystic philosopher, world renowned underground occult expert and prolific writer of over 90 books on systemology, ancient history, magic and "esoteric archaeology" since 1995. He founded Mardukite Ministries (Mardukite Zuism) in 2008, is director of Mardukite Research Organization (Mardukite Academy) and its New Thought division "The Systemology Society."

PUBLISHED BY THE **JOSHUA FREE** IMPRINT REPRESENTING

The Founding Church of Mardukite Zuism
& Mardukite Academy of Systemology

mardukite.com

www.ingramcontent.com/pod-product-compliance
Lightning Source LLC
Chambersburg PA
CBHW080749120626
46557CB00005B/1206